AF869067

Simulation as Grand Unification:

A Computational Framework for Physical and Metaphysical Coherence

By

Robert K. Restivo

Copyright © 2025 by Robert K. Restivo
All rights reserved. No part of this publication may be reproduced, distributed, or transmitted in any form or by any means, including photocopying, recording, or other electronic or mechanical methods, without the prior written permission of the publisher, except in the case of brief quotations embodied in critical reviews and certain other noncommercial uses permitted by copyright law.

For permission requests, write to:
Robert K. Restivo

robert@robertsrecordingreel.com

www.robertsrecordingreel.com

ISBNs

Hardcover: 9781971034003

Paperback: 9781971034010

Cover and interior design by Sora Graphics

Disclaimer:
This publication is for informational purposes only and does not constitute legal, financial, or business advice. Readers are encouraged to seek professional guidance tailored to their individual circumstances.

First Edition: December 2025
Published by Robert's Recording Reel
www.robertsrecordingreel.com

Printed on Earth

Master Table of Variables

Reference for symbols, constants, and variables used throughout the book.

1. Physical Constants

Symbol	Description	Interpretation / Context
c	Speed of light	Maximum update rate of spacetime "processor"
ħ	Reduced Planck constant	Quantum of action; sets micro-level scale
G	Gravitational constant	Strength of gravity; macro-level geometry rendering
k	Boltzmann constant	Links energy, temperature, and information
T	Temperature	Physical temperature or computational energy equivalent
E	Energy	Work done by or within the system
m	Mass	Matter represented as informational packets
Δx, Δt	Discrete space/time intervals	Planck-scale resolution units
ρ	Density	Mass-energy density; drives curvature/memory allocation
Φ	Gravitational potential	Macro-level field approximation

2. Simulation Variables

Symbol	Description	Interpretation / Context
F	Force	Output of physics "API" functions (gravity, EM)
LOD	Level of Detail	Computational rendering level; macro smoothing
Voxel	Volumetric pixel	Discrete unit of space-time memory allocation
CMB	Cosmic Microwave Background	Observed radiation; may contain compression artifacts
Axis of Evil	Directional anomaly in CMB	Possible alignment of early-universe rendering grid
Lossy Compression	Approximate data reduction	Optimization in early-universe simulation; causes artifacts
TORS	Theory of Rendered Simulation	Unifying framework: laws as rendering protocols
TOE	Theory of Everything	Traditional physics goal; replaced by TORS in this book

3. Information & Observer Variables

Symbol	Description	Interpretation / Context
ψ	Quantum wavefunction	Probabilistic representation of micro-states
$\Delta\psi$	Wavefunction collapse	Resolution when observation occurs

Symbol	Description	Interpretation / Context
S	Entropy	Measure of inaccessible information; side effect of computation
I	Information	Bit content; substrate of reality
χ	Observer state	Local rendering or conscious interface
τ	Processor tick / iteration	Fundamental unit of simulated time

Notes for Readers:

- Symbols may have **dual interpretations**: physical reality and computational analogy.
- Use this table to track concepts across **macro and micro scales**, **information theory**, and **observer effects**.
- Think of TORS variables as **the universe's "settings and subroutines"**, rather than inscrutable physics.

Prologue

A Moment of Realization

The cursor froze.

Not on the screen, but in the air — a bird mid-flight, wings extended, suddenly suspended like an image file that hadn't finished rendering. Its feathers trembled against a wind that no longer moved. For an instant, the world itself stuttered. The hum of traffic outside fell silent, the digital clock on his desk stilled mid-blink, and shadows ceased their crawl across the wall. Even his own breath hung heavy and unmoving in his chest, caught in a moment outside of time.

His eyes darted around, searching for motion, for any sign of life, but everything held that eerie, absolute stillness — the kind that only dreams achieve, when reality feels more like a painted backdrop than a living scene.

And then, just as suddenly, it all resumed. The bird darted onward. The clock blinked. The traffic roared back into being with the low, thunderous growl of engines. Sunlight poured once again across the room, catching dust motes that tumbled like sparks shaken from an unseen fire.

He rubbed his eyes, convinced he was overtired — another late night, another marathon at the keyboard. But the sensation gnawed at him, electric and alien, an itch under the skin of his mind. Not a dream. Not fatigue. Something else. A glitch.

He stood and pressed his palms against the window, feeling for the familiar vibrations of the street below: footsteps, engines, the low thrum of the city. They were there, but thin somehow, like a recording played back at half-volume.

If the universe were what it seemed — a smooth, continuous fabric governed by immutable laws — such a stutter should be impossible. Yet, here it was: reality behaving like a lagging program.

He sat back down, heart hammering. The room felt both utterly mundane and faintly artificial — as if every object had been hastily rendered just to fill his field of view. For the first time, the thought that had haunted philosophy and science fiction alike no longer seemed absurd.

What if this isn't the base layer?
What if the world is running on something deeper — a code, a system, a simulation?

He glanced at his screen. The cursor blinked, steady now, but he no longer trusted it. He no longer trusted *anything*.

It didn't feel like science fiction anymore. It felt like the beginning of an answer — and the end of something else.

Preface

Why This Is Being Written — And Why Now

This book was born out of a persistent disquiet — a tension that has lived for decades between what we know and what we *feel* we should know by now. For more than a century, physics has sought not merely to describe the world, but to *unify* it — to distill all the motions of planets and particles, all the forces of nature and the flickering probabilities of the quantum realm, into one coherent principle. The search for the Grand Unified Theory — that elusive synthesis of gravity and quantum mechanics — has become the holy grail of modern science.

And yet, despite a century of brilliance, the grail remains hidden. The deeper we drill into matter, the stranger it becomes. What once appeared as solid atoms now dissolve into fields of probability. The neat determinism of Newton gives way to a world that behaves as though it's waiting to be *asked* how it should exist. Even space and time — the very scaffolding upon which reality seems built — are revealed to be flexible, relative, and perhaps not fundamental at all.

It is as if we are probing the limits not merely of knowledge, but of *the system that makes knowledge possible.*

At the same time, culture itself has entered a parallel phase shift. Concepts that were once dismissed as metaphysical curiosities — the possibility of simulated worlds, the idea that reality might be computed, or that information itself might be the substrate of existence — have migrated from the fringes of philosophy into the language of physics. When Nobel laureates describe the universe as a “hologram,” when cosmologists speak of the “computational capacity” of the cosmos, when computer scientists find physical limits mirrored in information theory — it is clear that the intellectual terrain has changed.

The boundary between physics and metaphysics has blurred.

This book emerges from that blurred region — not as a manifesto, but as a map. It proceeds from the conviction that the **simulation hypothesis** is not merely an amusing philosophical parlor game, nor a technological fantasy born of Silicon Valley excess. It is, potentially, a *useful ontology*: a way to think coherently about the discontinuities, paradoxes, and apparent “patchwork” nature of modern physical theory.

The simulation framework, if treated seriously, offers a new way to reinterpret the contradictions that plague our models — to understand why gravity resists quantization, why constants appear fine-tuned, why observation seems to shape reality itself. It may also offer a bridge between the human and the cosmic — between the computational and the conscious — by reframing existence not as a collection of objects but as a flow of *information instantiated through observation.*

This is not, therefore, a work of idle speculation. It is a thought experiment in disciplined wonder.

It takes as its premise that we have reached an inflection point: one where physics, computation, and philosophy are converging on a shared mystery. If we continue to treat these domains as separate — if we keep physics to the laboratory and metaphysics to the armchair — we may never grasp the deeper structure underlying both.

And so, this book asks: What if the impasse of modern unification is not a failure of mathematics or measurement, but a failure of *ontology*? What if our theories are incomplete not because nature is elusive, but because we are asking the wrong kind of question?

To explore that possibility is not to abandon science — it is to extend its method into the realm it has always feared: the realm of *why*.

This is why the book is being written — and why now. Because the gap between our models and our intuitions is widening. Because the simulation framework, if taken seriously, may not only close that gap but transform it into a new foundation.

Because the time has come to treat speculative physics with the same seriousness that speculative fiction once treated destiny.

And because every civilization, at the height of its knowledge, eventually asks the same question in a different language:

What, exactly, is real?

Speculative but Disciplined

This is not a manifesto. It is not an article of faith, nor a claim that we are "definitely" living in a simulation. Rather, it is a disciplined attempt to explore what happens if we treat the simulation hypothesis with the same seriousness that physicists bring to quantum field theory, or that natural philosophers once brought to the idea of atoms centuries before they could be observed.

Speculation is not the enemy of science; undisciplined speculation is. The history of physics is a story of audacious ideas that were first ridiculed, then debated, then measured. At various points, relativity, black holes, and even the quantum itself were little more than bold conjectures — yet by subjecting them to clear reasoning and precise constraints, scientists transformed them into tools of understanding.

That is the spirit in which this book is written.

To explore the simulation hypothesis responsibly requires a rigorous self-awareness. Every speculative leap must be clearly marked as such. Every analogy to computation must be anchored, where possible, in known physics — whether information theory, thermodynamics, or the mathematical structures of field theory. Where grounding does not yet exist, the book will frame its claims as conjectures, not conclusions.

This is a map of possibilities, not a proclamation of truth.

The arguments herein will move in a deliberate rhythm: from conceptual frameworks to physical observations, from formal conjecture to potential tests. The reader will always be able to distinguish between what is established, what is plausible, and what is purely exploratory. In this way, the book seeks to model a kind of "ontological experimentation": bold enough to push the boundaries of current theory, but careful enough to keep speculation tethered to the real constraints of mathematics, logic, and empirical science.

If the simulation hypothesis has any chance of becoming more than a philosophical toy, it must pass through this crucible of disciplined speculation. This book is an invitation to do exactly that.

Who This Is For

This book is written for three kinds of readers — not defined by profession, but by temperament.

First, it is for the *curious physicist* who feels the subtle pull of unease — who looks at the unsolved seams between quantum mechanics and relativity and wonders whether the problem lies not in the math but in the metaphysics beneath it. You have mastered the tools of modern physics, yet suspect that something essential is missing from its foundations: a question that cannot be answered by more data alone. You are not content to treat anomaly as error; you see in anomaly the outline of a deeper pattern.

Second, it is for the *disillusioned philosopher* — the one who has watched metaphysics shrink under the triumph of empiricism, yet knows that science itself rests upon metaphysical commitments: to causality, to order, to intelligibility. You have seen philosophy retreat from the frontier it once shared with physics, and you sense that it is time for a reunion. For you, this book offers a bridge — a way to engage metaphysical speculation not as nostalgia, but as necessary complement.

Third, it is for the *open-minded skeptic* — the careful thinker who distrusts grand claims yet remains honest enough to admit that our current models of the universe, however successful, are patchwork. You may not believe in any ultimate theory, but you recognize the humility in asking whether what we call "reality" might itself be an interface: a functional illusion built for observers like us.

If you find yourself in any of these descriptions — or somewhere in between — this work is for you. It is not an argument to convert or a sermon to persuade. It is, instead, an invitation to *consider* — to let an alternative ontology take shape, if only as a thought experiment, and to see what physics looks like when the possibility of simulation is treated as a scientific hypothesis rather than a philosophical curiosity.

It asks for no belief. Only for attention — and for the willingness to momentarily set aside the assumption that the universe must be exactly what it appears to be.

Table of Contents

Introduction

The Grand Unified Illusion

The Quest for a Grand Unified Theory (GUT)

Since the dawn of modern science, the deepest impulse of physics has been unification — the dream that beneath the riot of appearances lies a single, elegant order. It is an aesthetic as much as a scientific pursuit, a belief that the universe, however vast or strange, must be comprehensible in principle.

In the seventeenth century, **Isaac Newton** achieved the first great unification. He showed that the same invisible force that pulled an apple to the ground also curved the path of the Moon. The heavens and the Earth were not governed by separate laws after all — there was only one. His *Principia* collapsed the boundary between celestial and terrestrial physics and gave humanity its first taste of a universe ruled by universal principles.

Two centuries later, **James Clerk Maxwell** performed a similar miracle. Electricity and magnetism, once thought distinct, were revealed to be two aspects of the same electromagnetic field. Maxwell's equations transformed not only physics but metaphysics: they implied that invisible fields could be as real as matter itself. Unification was not just possible — it was beautiful.

The twentieth century raised the stakes. **Albert Einstein** merged space and time into a single manifold: spacetime. Gravity, he realized, was not a force at all but the curvature of that manifold — geometry itself bending under the influence of mass and energy. Once again, two seemingly different things — motion and geometry — turned out to be one.

The pattern became a kind of myth of progress: wherever there were two, there must be one. The search for unity became physics' secular theology.

By the 1970s, this ambition took on a new name: the **Grand Unified Theory**, or GUT. The Standard Model had already succeeded in uniting electromagnetism and the weak nuclear force into a single "electroweak" interaction. The hope was that the strong nuclear force could be folded in next, and finally, gravity itself would join the chorus.

But gravity refused to cooperate. General Relativity — Einstein's continuum — spoke the language of smooth geometry, while quantum mechanics spoke in probabilities, quanta, and discontinuities. Each was breathtakingly successful in its domain, yet when brought together, their equations produced infinities, contradictions, and nonsense.

The result was an intellectual stalemate: a universe that behaves deterministically on the large scale and probabilistically on the small — a cosmos split between two incompatible languages.

By the 1990s, **string theory** appeared as a new contender, promising not just a GUT but a **Theory of Everything (TOE)** — a single mathematical structure in which all forces, all particles, and perhaps even spacetime itself, would emerge from the vibration of infinitesimal strings. It was bold, mathematically rich, and conceptually elegant. But decades later, the promised experimental confirmation has never come.

Instead, the more we refine our measurements, the more the anomalies multiply. The constants of nature — the fine-structure constant, the cosmological constant, the mass ratios of fundamental particles — appear tuned to improbable precision.

Quantum mechanics remains probabilistic at its core, as though hiding an algorithmic process beneath its surface. Gravity continues to resist quantization. And the "unified" theories of everything keep retreating into higher dimensions, more parameters, and more abstract mathematics — beautiful, but untestable.

In short, unification remains an illusion: a receding horizon that reveals the limits not just of our mathematics, but perhaps of our ontology — our very assumptions about what the universe *is*.

And so we arrive at a quiet but radical question — one that will guide the pages that follow:

> What if the inconsistency is not in the equations, but in the substrate itself?
>
> What if the universe is not a single, continuous physical fabric at all, but a *rendered environment* — a simulation whose apparent disunity arises from the way reality is computed rather than from what it is made of?

This, then, is where the story turns. The next part will explore how this idea — once dismissed as metaphysical fantasy — has evolved into a plausible framework for reinterpreting physics itself.

The Macro / Micro Divide: GR vs QM

At the heart of modern physics lies a deep and persistent fracture — a **schism between the very large and the very small**. On one side stands **General Relativity (GR)**, the majestic theory of spacetime geometry; on the other, **Quantum Mechanics (QM)**, the statistical dance of energy and uncertainty. Each reigns supreme within its own domain. Each collapses when asked to speak the other's language.

General Relativity is a theory of smoothness. It envisions the universe as a continuous, four-dimensional fabric — spacetime — warped and curved by the presence of mass and energy. Gravity, in this picture, is not a force transmitted through space but a manifestation of that curvature itself. Planets orbit stars not because they are "pulled," but because they follow the natural geodesics of this bent geometry. The equations of GR are deterministic and local: if you know the configuration of matter and energy at one moment, you can, in principle, calculate its evolution forever after. It is a theory of serene inevitability — elegant, geometric, and complete.

Quantum Mechanics, by contrast, is a theory of discreteness. It describes a world that is grainy, indeterminate, and probabilistic at its core. Energy comes in quanta, matter exists in superpositions, and the act of observation seems to collapse potentiality into actuality. The quantum realm is ruled not by geometry but by algebra — wavefunctions, operators, and probability amplitudes. It is *nonlocal*, meaning that correlations can appear to leap across space faster than light, as if the underlying substrate of reality were less like a field and more like a computation updating globally.

Both theories are stunningly accurate within their zones of applicability. GR predicts the bending of starlight, the precession of Mercury's orbit, and the expansion of the cosmos itself. QM explains the stability of atoms, the behavior of semiconductors, and the laser light guiding your mouse. Yet their conceptual foundations could hardly be more different.

When we try to merge them — to describe conditions where both gravity and quantum effects are strong, such as inside black holes or at the first instant after the Big Bang — the mathematics implodes. **Quantizing gravity** yields infinite values that cannot be renormalized. **Curving quantum fields** destroys their delicate coherence. Each theory presupposes a background — GR assumes a continuous manifold, QM assumes a probabilistic substrate — and those backgrounds are mutually incompatible.

For nearly a century, physicists have sought a bridge. String theory, loop quantum gravity, emergent spacetime models, holographic dualities — each promises reconciliation, and each remains incomplete. The more we refine our approaches, the more this divide begins to look **ontological rather than mathematical**. Perhaps it is not that our equations are wrong, but that our metaphors are misaligned.

If GR describes the macro-world as a continuous simulation of curvature, and QM describes the micro-world as discrete, probabilistic computation, then perhaps the incompatibility arises because they are **two renderings of the same underlying process**, seen at different levels of resolution.

From that perspective, gravity and quantum mechanics are not rival languages — they are complementary scripts, compiled differently for performance and scale. The "macro/micro divide" may not mark the boundary between two kinds of physics, but between two *modes of rendering reality*.

The Limits of Reductionism

For centuries, reductionism has been the guiding light of science — the idea that by taking things apart, we can ultimately understand the whole. The success stories of this approach are undeniable. Chemistry became comprehensible through the discovery of atoms; biology was revolutionized by the molecular understanding of DNA; and physics reached unprecedented explanatory power by revealing that protons and neutrons are built from quarks, which themselves may arise from vibrating strings. Each reduction seemed to peel back another layer of reality, bringing us closer to the core — to the ultimate building blocks of existence.

But this triumph carries an assumption that is rarely questioned: that there *is* a bottom layer — a most fundamental stratum beneath which there is nothing left to explain. This assumption has guided the scientific enterprise for centuries, shaping not only how we ask questions but what kinds of questions we even consider meaningful. Yet as our theories have deepened, that confidence has begun to waver.

Reductionism presumes that causality flows from the bottom up — that the properties of the parts determine the behavior of the whole. But increasingly, evidence suggests that the reverse may also be true: wholes can constrain parts. Systems can exhibit "emergent" behaviors — phenomena that cannot be derived simply by summing their constituents. The flocking of birds, the superconductivity of electrons, the consciousness arising from neural networks — each reveals that new rules emerge at higher levels of organization, rules that are not merely the sum of their microphysical details.

In computation, this is obvious. The binary switches of machine code do not explain the meaning of a poem generated on the screen. The poem is "real" in a different sense — an emergent property of the code but not reducible to it. The layers coexist, each valid within its own context, each indispensable to the system's functioning. Trying to reduce one to the other is not only impossible but misses the essence of what makes each layer meaningful.

Perhaps physics has reached a similar frontier. Quantum fields and spacetime curvature may not be different "levels" of the same hierarchy, but parallel abstractions — complementary descriptions of reality that only make sense within their respective domains. Gravity may not *emerge from* quantum mechanics, nor quantum mechanics *emerge from* gravity; instead, both might be expressions of a deeper informational or relational substrate that transcends both frameworks.

If so, then our century-long quest for a single, all-encompassing "theory of everything" may be chasing a mirage born of reductionist habit. The universe might not be a machine built from smaller and smaller gears. It might be more like a web of interdependent processes, where causation loops through multiple scales — where meaning and structure flow not only downward but upward, sideways, and across.

To move beyond reductionism, we may have to abandon the notion of "fundamentality" altogether — not to give up on understanding, but to redefine what understanding means. The truth might not lie at the bottom of the stack, but in the relationships that bind all layers together.

The Overlooked Hypothesis: A Fault in Ontology

When physicists confront contradictions between two successful theories, the instinct is almost always the same: look for a mathematical mistake. Perhaps the equations are incomplete, or the constants need to be tuned, or a hidden variable has been overlooked. The working assumption is that the *framework* is sound; only the details need refinement.

But what if that assumption itself is the problem? What if the mathematics is perfectly correct within its own terms — and the error lies not in the equations but in the *ontology* that underpins them, the picture of reality we take for granted?

General relativity treats spacetime as a smooth continuum, an infinite fabric that bends under mass and energy. Quantum mechanics, by contrast, treats reality as granular, probabilistic, discontinuous. Both theories still smuggle in a shared assumption: that "the universe" is an objective thing "out there," a pre-existing stage on which events unfold. Even when we question determinism or locality, we rarely question that stage itself.

But what if there is no stage? What if the universe is not an object at all but a process — a ceaseless act of becoming rather than a thing that *is*? In such a view, the contradictions between relativity and quantum theory may not be the result of incomplete unification but of category error. We may be trying to reconcile two descriptions that were never about the same "stuff" to begin with.

Consider how a video game works. The images on the screen appear continuous and fully realized, but in reality they are rendered on the fly, frame by frame, based on the player's perspective and the game engine's rules. The "world" inside the game does not exist as a complete, precomputed object. It is an emergent phenomenon of code, data, and processing power — a process masquerading as a place.

This is not merely an analogy. In computation and information theory, such "on-demand rendering" is normal. In biology, too, we find processes rather than static things: a living cell is not a fixed object but a dynamic network of interactions continually renewing itself. Why should the cosmos be any different?

From this vantage, the elusive "Grand Unified Theory" may be elusive because it does not exist in the way we imagine. We are searching for a single framework that unites space, time, matter, and energy, assuming these are fundamental entities. But if they are emergent — if they are outputs of a deeper informational substrate — then our equations are like trying to unify the pixels and polygons of a game without acknowledging the code that generates them.

This book begins from that hypothesis. Not as a dogma or an article of faith, but as an invitation to test a radical reframing: that the true unification problem is not about forces or particles, but about ontology. We may need to shift from describing what the universe *is* to understanding what the universe *does*. Only then might the contradictions dissolve, not because we've forced incompatible equations together, but because we've stepped outside the illusion they both inhabit.

Part I

The Case for Simulation as a Paradigm

Before any framework can be evaluated as a serious candidate for unification, it must first be allowed to stand as a paradigm. Throughout the history of science, ideas that later became foundational began as outliers. The heliocentric model was once heresy; the germ theory of disease was once "miasma denial." Quantum mechanics itself began as an awkward patch to explain blackbody radiation. Paradigms do not emerge fully formed from the mainstream; they germinate at the margins, as irritants to the prevailing order of thought, until the anomalies they explain can no longer be ignored.

The simulation hypothesis has, until recently, occupied precisely that liminal space — somewhere between philosophy seminar and late-night speculation, more the province of science fiction writers than working physicists. And yet, in the past two decades, the intellectual terrain has shifted. Nobel Prize winners now describe the universe as a "hologram." Cosmologists speak in the language of computation and information theory. Experimental physicists design tests for "quantum error correction" in spacetime. What was once a fringe conjecture has begun to acquire a scientifically inflected vocabulary.

The central claim of this Part is straightforward but disruptive: if we are to understand why the Grand Unified Theory has remained elusive for over a century, we must consider the possibility that the impasse is not mathematical but ontological. The very picture of the universe as a single, continuous, pre-existing physical "thing" may itself be the obstacle. Perhaps, like software running on hardware, what we experience as "reality" is more fruitfully modeled as a *process* — a dynamic, algorithmic unfolding that generates the appearance of physical law only as needed, when and where it is required.

This shift of framing has consequences. It reorients the unification problem from one of merging incompatible forces to one of identifying the underlying "rendering engine" beneath them all. Just as a video game's graphics and physics appear seamless only because of a deeper code running invisibly in the background, so too might space, time, matter, and energy be emergent phenomena of an informational substrate.

This section develops the case in two steps.

First, we revisit the simulation hypothesis itself, tracing it from Nick Bostrom's original philosophical formulation to its present-day scientific incarnations — from digital physics to holographic principles and quantum computation. Along the way, we will show how a hypothesis that once seemed speculative now resonates with mainstream trends in theoretical physics.

Second, we examine how certain stubborn features of our universe — the observer effect, the quantization of action, the speed of light as a seemingly arbitrary limit — align more naturally when viewed through the lens of a "rendered reality" rather than a pre-existing one.

Taken together, these moves do not "prove" the simulation hypothesis. Rather, they open up a disciplined space for considering it as a working paradigm — not as a metaphysical curiosity, but as a framework that may illuminate the deepest puzzles of modern physics.

Chapter 1

The Simulation Hypothesis Revisiting the Core Argument

Bostrom's Trilemma

In 2003, philosopher Nick Bostrom published a paper in *Philosophical Quarterly* that would ripple far beyond the academic world. It offered no equations, no cosmological data, no new physical theory — only a logical structure. Yet that structure, now known as "Bostrom's trilemma," captured the public imagination in a way few philosophical arguments ever have. It was simple, elegant, and unsettling.

The trilemma can be stated as follows:

1. **Almost all civilizations at our level of development go extinct before reaching technological maturity.**
2. **Almost no advanced civilizations are interested in running simulations of conscious beings.**
3. **We are almost certainly living in a simulation.**

The argument is not a straightforward claim but a forced choice. At least one of these three statements must be true. In fact, Bostrom's analysis suggests that one and only one is likely to dominate. If the first is true — that most civilizations self-destruct before developing the capability to simulate entire worlds — then there will be very few, if any, simulations. If the second is true — that even advanced civilizations rarely choose to simulate conscious beings — then again, simulations are rare. But if both the first and second are false, then the third becomes virtually inevitable: we are almost certainly living inside a simulation.

Bostrom's point was probabilistic, not metaphysical. He wasn't claiming to know *which* of the three is correct. Instead, he was highlighting how lopsided the probabilities become under plausible assumptions. If advanced civilizations survive and have even a modest inclination to run ancestor simulations — digital reconstructions of their past, training grounds for AI, or experimental sandboxes — then the number of simulated conscious beings would vastly outnumber the "base reality" beings. Statistically, any given conscious observer would be overwhelmingly likely to be among the simulated rather than the simulators.

This reasoning has the stark, unavoidable flavor of actuarial math: you don't have to like the conclusion to accept the odds. And that, perhaps, is why the idea caught fire. Unlike older versions of the "illusion" hypothesis — dreams within dreams, brains in vats, or Descartes's demon — Bostrom's trilemma rests not on speculation about the nature of consciousness but on trends in technology and the logic of large numbers.

Yet, two decades later, the trilemma is often misunderstood. It is not an argument *for* the simulation hypothesis in the way the Big Bang theory is an argument *for* cosmic expansion. It is an argument for the *conditional plausibility* of simulation as a paradigm. If you reject it, you must reject it at one of the three branches: either civilizations do not survive, they do not simulate, or we are simulated. There is no neutral "opt-out" position.

For our purposes, Bostrom's trilemma serves as a starting point, not an endpoint. This book is not about debating whether we are simulated in the philosophical sense but about asking a deeper question: **if** the simulation hypothesis were true, how would it help us understand the impasse in modern physics? How would it recast the problem of unification — and perhaps, finally, give us a way to move beyond it?

Evolution of the Hypothesis Post-Bostrom

When Bostrom first presented his trilemma in 2003, it landed somewhere between philosophy and provocation — an elegant thought experiment that bridged the gap between metaphysics and mathematics. But the years since have transformed that thought experiment into something more substantial. What began as a philosophical curiosity has begun to merge with developments in computer science, physics, and cosmology. The conversation, once abstract, is now empirical — or at least empirically adjacent.

Information as a Physical Quantity

One of the most significant shifts came not from philosophy but from physics itself: the rise of **information theory** as a central pillar of the natural sciences. The recognition that *information* is not merely a descriptor of reality but an *ingredient* of it has changed how physicists think about everything from entropy to black holes. Claude Shannon's foundational insight — that information can be quantified in bits — found new resonance in thermodynamics through the work of Rolf Landauer, who showed that information processing has an unavoidable energy cost. Later, physicists like John Archibald Wheeler went further, proposing his famous aphorism: *"It from bit"* — the idea that every physical "it," every particle or field, ultimately derives from an informational "bit."

Quantum mechanics has only deepened this insight. The quantum bit, or *qubit*, can exist in a superposition of states, encoding probabilities rather than certainties. When measured, the wavefunction "collapses," and a definite value appears — not unlike the way a rendering engine produces a visible frame when data is queried. Thus, what began as an analogy between information and matter has evolved into a literal equivalence in many modern interpretations: reality, at its foundation, is informational.

The Rise of Digital Physics

From this realization emerged what is now known as **digital physics** — the hypothesis that the universe itself operates as a kind of computation. Edward Fredkin, one of its early proponents, proposed in the 1980s that the universe could be modeled as a *cellular automaton*: a vast network of discrete cells, each updating according to simple rules, producing the complex patterns we perceive as physical law. Stephen Wolfram's later work in *A New Kind of Science* extended this idea, arguing that the richness of nature may arise not from continuous equations but from simple computational rules iterated at unimaginable scales.

Seth Lloyd took the argument further still, estimating that the universe — if treated as a quantum computer — has performed approximately 10^{90} logical operations since the Big Bang. This isn't a metaphor. In Lloyd's formulation, *the universe literally computes its own evolution.* Every interaction, every entanglement, every exchange of energy is an update to a cosmic algorithm. Under this lens, what we call "physical law" becomes the operating system of existence itself.

The Holographic Turn

Meanwhile, developments in theoretical physics — particularly in quantum gravity and string theory — began to echo these computational intuitions. The **Holographic Principle**, first articulated by Gerard 't Hooft and later developed by Leonard Susskind, proposes that all the information contained within a volume of space can be represented on its boundary, just as a hologram encodes three-dimensional information on a two-dimensional surface. In the celebrated AdS/CFT correspondence formulated by Juan Maldacena, an entire higher-dimensional universe (Anti-de Sitter space) is mathematically equivalent to a lower-dimensional conformal field theory on its edge.

This is not just elegant mathematics; it is compression — precisely the kind of encoding logic familiar to computer science. A three-dimensional reality "rendered" from two-dimensional data suggests that the universe itself might operate on principles strikingly similar to data storage and retrieval. To many physicists, this is coincidence. To others, it hints that the "hardware" of the cosmos is informational by nature.

From Speculation to Framework

Taken together, these threads — information theory, digital physics, and the holographic principle — have begun to weave a new tapestry. What once sounded like a philosophical *what if* now reads as a plausible *how it might work.* The simulation hypothesis has thus evolved from metaphysical musing to conceptual framework, one that unites disparate insights across disciplines. It offers a new way to think about old problems: quantum indeterminacy, the observer effect, the fine-tuning of constants, and the stubborn incompatibility of General Relativity and Quantum Mechanics.

In short, what was once dismissed as science fiction is now increasingly recognized as a serious contender — not necessarily as *truth*, but as a powerful model. Like all good scientific paradigms, its value lies not in whether it is "right," but in whether it helps us make sense of what otherwise seems incoherent.

From Metaphysics to Scientific Proposal

For much of the twentieth century, the claim that "the universe is like a computer" would have been filed under poetic metaphor or speculative philosophy. Today it no longer sounds far-fetched. The language of computation now permeates mainstream physics, not because it is fashionable but because the mathematics keeps pointing in that direction.

Physicists such as **John Archibald Wheeler**, who coined the phrase *"it from bit,"* argued that information is not merely a convenient description but the substrate from which physical reality emerges. "Every particle, every field, even the spacetime continuum itself," Wheeler wrote, "derives its function, its meaning, its very existence entirely from binary choices, bits." What began as a bold slogan has matured into a research agenda: if information is primary, then matter and energy are secondary — emergent rather than fundamental.

Even more provocatively, **James Gates**, working on supersymmetry, discovered that certain equations seemed to encode structures identical to computer **error-correcting codes** — the same kind used in digital communications to detect and correct corrupted data. Gates himself is careful in how he frames this finding, but the resonance is undeniable: why should the deep symmetries of nature contain patterns indistinguishable from those we engineer to stabilize information? Unless, perhaps, nature itself is stabilizing information.

This is the subtle but crucial shift. The simulation hypothesis, once a **philosophical provocation**, has become a **scientific proposal.** It is no longer an idle "what if" about metaphysics; it is a framework that can be evaluated by the same criteria as any other scientific model.

- **Does it organize known data?** A good theory simplifies, unifies, and clarifies. The simulation hypothesis offers a new way to interpret phenomena such as the observer effect, Planck-scale discreteness, and the fine-tuning of constants — not as unrelated quirks but as signatures of a deeper process.
- **Does it dissolve anomalies?** Instead of treating the incompatibility of General Relativity and Quantum Mechanics as a bug, the simulation framework treats it as a clue. Local, deterministic rules at the macroscopic level and probabilistic, nonlocal rules at the microscopic level are exactly what one might expect from a layered computational system.
- **Does it generate testable predictions?** Perhaps most importantly, the hypothesis is beginning to inspire experimental proposals — from detecting lattice-like anisotropies in cosmic rays to probing whether quantum randomness might show subtle "pseudorandom" structure. These tests may be crude at first, but they mark a transition from speculation to empiricism.

To call this a "scientific proposal" does not mean the simulation hypothesis is proven, or even necessarily correct. It means only that it has matured to the point where it can be examined with the tools of science: hypothesis, model, prediction, falsification. In that sense, it stands exactly where atomic theory once stood in the seventeenth century — a speculative but disciplined attempt to make sense of a world whose visible behavior hints at a hidden structure beneath.

Chapter 2

The Physics of a Rendered Reality

If the simulation hypothesis is to stand as more than metaphor or thought experiment, it must find purchase in physics itself — in the very structure of the natural world. It cannot remain the province of philosophy alone. The claim that we inhabit a rendered or computational universe is, at its heart, a claim about *how physical law operates.*

Remarkably, many of the strangest and most persistent mysteries in modern physics — quantum indeterminacy, the discreteness of energy levels, the relativistic constraints on information transfer, the holographic principle, and even the fine-tuning of constants — begin to align when examined through the lens of computational ontology. What once appeared as separate puzzles start to rhyme, each suggesting that the universe behaves as though it were optimizing for informational efficiency.

A rendered reality framework does not replace physics. Rather, it offers a meta-interpretation — a unifying language that connects quantum theory, relativity, and information science under a single principle: that *existence itself may be the dynamic output of an information-processing substrate.* The simulation is not a metaphor for the universe. It may be what the universe *is doing.*

This chapter explores that possibility by reinterpreting key phenomena in light of rendering logic. We begin with the most striking of all: the role of the observer in determining physical reality.

Observer Effects and Quantum Indeterminacy

At the heart of quantum mechanics lies a paradox. Particles — the fundamental building blocks of the physical world — exist not as definite things but as probability distributions, "wavefunctions" describing all possible states simultaneously. Only when observed do they appear to "choose" a definite outcome. This transition, from possibility to actuality, is known as *wavefunction collapse.*

For nearly a century, physicists have struggled to make sense of this observer effect. Does consciousness cause collapse, as some interpretations suggest? Is it a matter of decoherence — the statistical loss of superposition when a quantum system interacts with its environment? Or is there something deeper, something in the nature of existence itself that makes the universe *responsive* rather than *absolute*?

In a rendered reality framework, the observer effect is not a problem to be solved, but a feature to be expected. Rendering engines in computer graphics do not calculate every pixel, texture, and polygon in a world simultaneously. They render only what is required — what falls within the player's field of view, what is relevant to the camera's perspective, or what will soon come into causal contact with it. Everything else remains in an unrendered, potential state — existing as stored data, not as actualized form.

This principle of *deferred rendering* is computationally efficient, conserving processing power while maintaining a seamless experience of continuity. Transposed into the language of physics, the "collapse" of the wavefunction is nothing other than the universe rendering a definite state when — and only when — an observer's interaction requires it.

This interpretation reframes the observer effect not as an act of metaphysical magic or subjective intrusion into objective reality, but as an expression of a universal economy of information. It suggests that the universe itself operates under constraints of optimization — minimizing unnecessary computation, conserving energy, and generating only what is needed for consistent causal coherence.

The implications of this are profound. It means that quantum uncertainty is not a flaw or gap in our understanding of nature, but the signature of a system that calculates reality on demand. It transforms the question "Why does observation matter?" into "Why would a simulated universe need to compute what no one is observing?"

From this perspective, indeterminacy is not chaos — it is deferred decision. The universe, like a rendering engine, carries all potential outcomes in latent form until observation collapses them into actuality. What we perceive as randomness may, in fact, be the surface behavior of a deterministic — or at least rule-bound — underlying code, optimizing for experiential continuity rather than exhaustive precomputation.

Thus, the observer effect becomes a clue — perhaps the most important one — that reality operates as an information-efficient system. And once that door is opened, the parallels between physics and computation only deepen: space-time as a memory grid, energy as processing cost, entropy as information loss, and even gravity as an emergent distortion in data density.

But before those analogies can be meaningfully developed, one must understand how deeply quantum theory already assumes an informational substrate — a topic we turn to next.

The Speed of Light as the Processor's Tick Rate

Among all the constants in nature, none is more fundamental — or more mysterious — than the speed of light. It is not merely the velocity of photons, but the upper bound for the transmission of *any* information through space-time. No object, no signal, no causal influence can exceed it. To the physicist, this constant, c, is woven into the very fabric of relativity — defining the relationship between space, time, and energy. But to the philosopher of simulation, it may signify something else entirely: **the tick rate of the universal processor.**

If the universe is a computed environment, then every event — every interaction, every shift in energy, every quantum transition — represents a discrete update in the underlying information state. Reality would proceed not continuously, but in infinitesimal computational steps, each governed by a maximum refresh rate. In such a system, the speed of light is not an arbitrary constant decreed by nature, but the maximum speed at which information can be propagated within the simulation's architecture — *the system clock's ultimate frequency.*

The Universal Clock

In digital systems, the processor's *clock rate* determines how fast information can be processed, transferred, and synchronized. No signal can outpace the system's fundamental timing cycle, because every instruction — every update — depends on it. To exceed the tick rate would mean to access a state that does not yet exist in the sequence of computations; it would be akin to skipping frames in a rendered simulation before those frames have been calculated.

By analogy, light speed may define the cadence by which our universe updates its state vector — the "refresh interval" of physical reality. When an event occurs, the propagation of that information through space-time cannot exceed the system's intrinsic rhythm. Hence, c becomes not just a number, but a boundary condition on causality itself: a computational synchronization limit that maintains coherence across the simulation.

This interpretation gives an intuitive rationale for the *universality* of light speed. It is not the property of photons per se, but of the substrate through which all information flows — the shared timing protocol of reality's operating system.

Relativistic Effects as Optimization Behavior

In Einstein's relativity, objects approaching the speed of light experience *time dilation* and *length contraction*: time slows down and distances shrink along the direction of travel. From within the simulation framework, these effects are not strange at all — they are signs of **optimization under computational load.**

As an object's velocity increases, the system must update its state more frequently to maintain causal coherence across frames. Near the threshold of c, the computational demand approaches infinity: each additional increment of velocity requires exponentially more processing power to keep the simulation consistent. The universe's solution is elegant: *it slows time down* from the perspective of the moving frame, effectively reducing the rate at which updates are required for that observer.

Length contraction, likewise, minimizes the spatial region that must be dynamically maintained. Both phenomena conserve computational resources, preserving the internal coherence of the system without exceeding the tick rate. In the same way that a graphics engine dynamically lowers rendering resolution when the frame rate drops — prioritizing continuity of motion over perfect detail — the universe adjusts space and time to preserve consistency within its finite processing bandwidth.

These relativistic adjustments can therefore be read as *adaptive optimizations* of a cosmic rendering system: elegant compromises between fidelity and performance.

A Constraint of System Architecture

When viewed through this lens, the speed of light is no longer an inexplicable brute fact, but a **design constraint** — an architectural boundary necessary for stability and causality. It ensures that no event can outpace the system's synchronization mechanism, preventing paradoxes such as effects preceding causes.

This recontextualizes relativity itself. The constancy of c across all reference frames — once regarded as one of the great mysteries of 20th-century physics — becomes a natural and even necessary feature of a discretized, computational universe. Every observer experiences the same speed limit because every process in the simulation is governed by the same global tick rate.

The profound insight here is that what we have taken as immutable law may in fact be *engineering*. The universe, like any well-designed system, preserves stability through constraint. The light-speed limit may be the safety margin of cosmic computation, preventing informational overflow and maintaining the causal fabric that allows reality to unfold coherently.

Toward a Computational Relativity

If this interpretation holds even partially true, it suggests a new direction for physics: a **computational reinterpretation of relativity.** The next frontier would be to identify how known relativistic quantities — such as energy, mass, and momentum — correspond to processing cost, data density, and computational throughput. Already, some theories point in this direction: the relationship $E=mc^2$ can be understood as a conversion factor between informational states (mass as stored data) and dynamic computation (energy as processing activity).

The simulation hypothesis, then, does not seek to discard Einstein's framework but to embed it in a deeper one — one that recognizes that space-time geometry and informational architecture are two expressions of the same thing.

What physics describes as the curvature of space and the flow of time may ultimately be the byproducts of a universal computation running at a finite tick rate, where the speed of light marks the rhythm of existence itself.

The Planck Scale as Pixel Resolution

At the foundation of modern physics lie two quantities that seem to mark the outermost edges of meaning: the **Planck length** (1.616×10^{-35} meters) and the **Planck time** (5.39×10^{-44} seconds). Below these scales, our mathematical descriptions lose coherence. The very notions of "distance" and "duration" cease to apply. Attempts to peer beneath them yield not sharper focus but conceptual blur — as if the microscope of theory itself reaches a hard limit of magnification.

Traditionally, physicists interpret these limits as natural consequences of combining quantum mechanics and relativity.

The Planck length, for example, represents the scale at which a photon's energy is so concentrated that it would curve spacetime into a black hole if one tried to measure anything smaller. It is not that reality stops there, but that our instruments — physical or mathematical — fail. Yet within the simulation paradigm, this failure may be telling us something more profound: that **there is nothing beneath the pixel.**

Discreteness as a Feature, Not a Flaw

In a rendered or computed universe, the Planck limits would correspond to the **native resolution** of reality's engine — the smallest meaningful "step" in the simulation's coordinate grid. Just as a digital image cannot be resolved below the scale of its individual pixels, the universe cannot be subdivided beyond its own informational granularity.

Space and time, under this view, are not continuous fabrics to be stretched infinitely thin, but **discrete lattices** — quantized data structures over which the physical "render" is computed. Every point in space, every moment in time, represents an address in that grid, updated according to local computational rules. Below that level, the concept of "between" no longer applies because there are no intermediate states to compute.

This reframing dissolves one of the thorniest problems in theoretical physics: why do attempts to merge quantum mechanics (which quantizes energy) with general relativity (which treats spacetime as smooth) lead to mathematical infinities? The answer may be that the underlying system is already quantized — but **not continuously deformable** in the way classical geometry assumes. The infinities arise not from nature's chaos, but from our insistence on describing a discretized system with continuous mathematics.

The Pixel Analogy and Its Implications

Imagine zooming into a high-resolution image. At first, it appears continuous: lines curve smoothly, gradients blend seamlessly. But magnify it enough, and its discrete structure emerges. Pixels, once imperceptible, reveal the fundamental limits of resolution. No amount of zoom can uncover "sub-pixels," because they simply do not exist in the data.

The same logic may govern the Planck frontier. Our universe, from the cosmic scale to the quantum, might be the visual output of a process that renders only as much fidelity as necessary for consistency at the observer's level of interaction. The smoothness of spacetime, like the apparent continuity of a photograph, is an emergent illusion — a byproduct of limited resolution observed from within the render.

From this perspective, Planck units acquire new meaning. They are not just awkward mathematical thresholds but **design parameters**: the minimum pixel size and frame duration that ensure coherence, stability, and sufficient granularity for emergent complexity. They are the constants that balance performance and fidelity in the architecture of being.

Why Quantizing Gravity Fails

For decades, physicists have sought to "quantize" gravity — to describe spacetime curvature in the same probabilistic language that governs particles and forces. Yet the enterprise continually breaks down at precisely the Planck scale, where the smooth geometry of relativity collides with the granular uncertainty of quantum theory.

In a rendered-reality framework, this persistent failure is diagnostic: it suggests that gravity is not a "force" acting within spacetime, but a property of the underlying grid itself — the way the system allocates processing resources to maintain the continuity of rendered objects. To quantize gravity would therefore be like trying to quantize the pixels of a display: the pixels *are* the quantization.

If spacetime is already discrete, then our unification problem may not lie in reconciling incompatible equations, but in **misunderstanding the level of description.** Quantum mechanics operates at the level of the system's data — probabilistic state transitions within the computational substrate. General relativity operates at the level of the rendered geometry — the emergent visual layer. Trying to merge them mathematically may be like trying to fuse source code with the imagery it generates: they describe different tiers of the same system.

The Planck Frontier as the Boundary of Knowability

The Planck scale, then, might not represent a mere technical limit of measurement but a **philosophical boundary** — the point at which ontology itself gives way to epistemology. Below that threshold, we are no longer describing what "is," but what "computes." Our instruments, no matter how refined, cannot detect sub-Planckian reality because it is not *rendered* for detection.

This does not mean such deeper levels are meaningless — only that they belong to the realm of source code, not perceptual space. The Planck constants delineate the line between *world* and *engine*, between the phenomena we inhabit and the substrate that makes them possible.

In this light, the Planck scale ceases to be an arbitrary mathematical artifact and becomes something far more profound:

a clue to the architecture of the cosmos — the grain of the simulation's fabric, the pixel grid on which existence is drawn.

Conservation Laws as Runtime Constraints

Every physics student learns early that certain quantities — energy, momentum, charge — are sacred. They can be exchanged, transformed, redistributed, but never destroyed or created from nothing. Conservation laws form the backbone of our physical intuition: they are the scaffolding upon which every successful theory rests. Yet rarely do we pause to ask the deeper question: *why* must they hold?

Traditionally, the answer is mathematical elegance. Emmy Noether's profound theorem of 1915 tells us that every continuous symmetry in nature corresponds to a conservation law. Time-translation symmetry gives rise to conservation of energy; spatial translation yields conservation of momentum; rotational symmetry produces conservation of angular momentum. The beauty of Noether's insight is that it links invariance to permanence: if the laws of physics do not change from one time or place to another, then certain quantities remain constant.

But Noether's theorem, as powerful as it is, describes *how* conservation arises — not *why* the universe possesses those symmetries in the first place. Why should reality obey such perfect invariance? Why should there exist global quantities that remain exactly balanced, even across cosmic timescales? From a simulation or computational standpoint, the answer takes on a new and striking clarity.

The Logic of Stability

In computation, invariants are not accidents; they are *enforced conditions* that prevent a system from diverging into incoherence.

Every complex program relies on runtime constraints — rules that maintain consistency across operations. For example, in a physics engine, momentum conservation ensures that simulated objects behave realistically even under limited computational resources. Without such constraints, small rounding errors would quickly amplify, leading to runaway behaviors or system crashes.

In this light, conservation laws in physics can be reinterpreted as **runtime enforcement mechanisms** within the universe's underlying code. They are not arbitrary metaphysical principles but **stability protocols**, guaranteeing that the system's internal state remains coherent as it evolves through time. The "law of conservation of energy" may be less an ontological truth about the universe's nature and more a structural requirement of a computational system that must balance its state variables perfectly to avoid drift.

Enforced Symmetry as Code Optimization

Seen through this lens, Noether's theorem reads like a developer's design document: every symmetry represents a deliberate optimization. If the simulation's physics engine treats time-translation as invariant — meaning the same code governs each update cycle — then energy must be conserved automatically. The same logic applies to space and momentum, to rotation and angular momentum.

These conservation rules are not "mystical absolutes," but byproducts of an elegant architecture. They reflect a balance between computational simplicity and representational fidelity: the universe's engine minimizes complexity by encoding physical evolution through symmetrical update rules. The result is a cosmos that appears law-bound not because it is "obliged" to be, but because symmetry is the most efficient way to generate consistency.

The Pragmatics of Enforcement

In computer simulations, enforcing conservation is not optional — it is essential for *runtime integrity*. A simulation without constraints would rapidly accumulate rounding errors, energy leaks, and inconsistencies that destroy realism. The same may be true of the cosmos. If physical laws did not strictly conserve energy, for example, the cumulative errors would propagate like bugs in the code, leading to unpredictable instabilities at macroscopic scales.

Thus, conservation laws can be viewed as **checksum protocols** — global conditions continually verified by the system to ensure coherence. Violations, when they appear (as in quantum uncertainty or vacuum fluctuations), are temporary local imbalances that resolve when integrated across the whole. From a computational standpoint, they are not exceptions to the rule but manifestations of rounding or sampling at the edge of precision — deviations that average to zero over sufficient iterations.

A Universe That Balances Its Ledger

This interpretation also reframes the metaphysical tone of physical law. The universe is not a passive arena where quantities merely happen to balance; it is an **active process of bookkeeping**, a self-consistent engine that perpetually reconciles its own accounts. The total energy of the cosmos, so far as we can measure, remains zero when one includes gravitational potential. This remarkable symmetry may not be coincidence — it may be the universe's way of ensuring that its computation remains netneutral, conserving informational and energetic balance across every update cycle.

In that sense, conservation is not a side effect of symmetry but its purpose. The cosmos does not merely obey mathematical constraints; it *runs on them*. The balance of energy and momentum is what allows the simulation to persist — not as a frozen tableau but as a living, self-correcting computation.

Beyond Law: Conservation as Design Philosophy

When viewed this way, the so-called "laws" of conservation take on a different philosophical texture. They are not inviolable decrees of some cosmic legislature, but **runtime strategies** embedded in the architecture of being. Their perfection is the perfection of good code — efficient, symmetrical, elegant, and minimal.

This shift does not demote physics; it elevates it. It invites us to treat the universe not as a set of detached equations describing a static thing, but as a dynamic system maintaining its own internal coherence through the continuous enforcement of constraints. It is not that conservation laws were "decided" by some external programmer, but that any self-consistent system capable of sustaining complexity over time must, by necessity, enforce conservation.

In short: conservation is the price of coherence. Stability demands balance, and balance demands rules.

Inverse Square Laws and Curved Space as Geometry Engines

One of the most striking regularities in physics is the **inverse-square law**. Gravity, electromagnetism, and even the intensity of light all fall off with the square of distance from a source. At first glance, this appears as a fundamental feature of our universe: forces diminish predictably as you move away from the origin. Yet within a simulation framework, the inverse-square law can be interpreted not as a primary law of nature, but as an **emergent property of the engine's geometry**.

Forces on a Discrete Grid

Imagine the universe as a three-dimensional lattice, a grid of points through which energy, information, and force propagate. Any "force" applied at one node must distribute across neighboring nodes, then their neighbors, and so on, forming a geometric gradient. The farther a point is from the source, the more nodes the effect must traverse, naturally diluting the strength according to the square of the distance — exactly what the inverse-square law describes.

From this perspective, the law is not a metaphysical dictate but a **consequence of computational architecture**. The engine does not "know" about gravity or electromagnetism intrinsically; it enforces consistency by propagating influence across a discrete spatial framework. What we measure as a $1/r^2$ falloff is simply the pattern that emerges from maintaining smoothness and coherence on a finite grid.

Curved Space as Projection Logic

General relativity offers a complementary description: mass and energy curve spacetime, and objects move along geodesics in this curved geometry. In simulation terms, curved space is a **rendering artifact of the engine's continuity algorithm**. To maintain smooth transitions between nodes and avoid jarring discontinuities, the system computes a curvature that ensures forces, motion, and light propagation appear continuous at all scales.

In other words, curvature is a **projection tool**. Just as a graphics engine interpolates textures over polygons to produce a smooth surface, the universe's rendering engine warps local geometry so that even at macroscopic scales, the discrete lattice behaves like a seamless, flowing manifold.

Two Sides of the Same Engine

Viewed together, the inverse-square law and curved space are two perspectives on the same computational process:

- **Inverse-square laws** reveal the discrete, node-to-node propagation of interactions across a 3D grid.
- **Curved space** reveals the engine's macro-level strategy for preserving continuity and smoothness across that grid.

Neither is more "real" than the other. Both are **descriptions of the rendering logic**, artifacts of a system designed to project a coherent, lawful reality from underlying discrete operations. The phenomena we call forces and curvature are not raw ingredients of the universe; they are **symptoms of the architecture** that generates the experience of reality.

Implications for Unification

Recognizing inverse-square laws and curvature as outputs of a geometry engine recasts the challenge of unification. Physicists have long tried to reconcile quantum fields with general relativity as if they were describing the same substance at different scales. But if one is a local propagation effect (forces across nodes) and the other a global continuity strategy (curvature projection), the difficulty is not mathematical error but a **category mismatch**.

Unification, then, may require understanding the **underlying engine** rather than forcing disparate laws into a single formula.

Toward a Unified Rendering Model

Individually, these reinterpretations may seem like clever metaphors — narrative tricks for making counterintuitive physics sound more familiar. Yet taken together, a striking pattern emerges: these phenomena begin to look less like isolated puzzles and more like coordinated features of an underlying architecture.

- **Quantum weirdness becomes deferred rendering.** The wavefunction doesn't "collapse" mysteriously; it's resolved only when needed, like a graphics engine calculating a texture only once the player approaches it.
- **Relativity becomes processor constraint.** The universal speed limit and its accompanying distortions (time dilation, length contraction) are no longer inexplicable absolutes but the performance envelope of the system, like clock speed in a CPU.
- **Planck limits become pixel resolution.** The smallest possible units of space and time are not arbitrary constants but the grid spacing of the engine — the point beyond which no finer resolution is meaningful or computable.
- **Conservation laws become runtime stability rules.** Energy, momentum, and charge are not metaphysically mandated but pragmatically enforced to keep the simulation coherent, like global invariants in a program.
- **Forces and curvature become geometry engines.** What we call "fields" and "spacetime curvature" may be the engine's way of distributing interactions and preserving

continuity across scales, just as a rendering system projects 3D geometry from encoded data.

In this light, the great divide between quantum mechanics and general relativity is reframed. They are not two contradictory descriptions of one "thing," but two modules of a larger codebase: the quantum engine handling deferred events at small scales, the relativistic engine handling continuous structure at large scales. Their apparent incompatibility may be like trying to run two subsystems written for different hardware architectures without recognizing the unifying operating system beneath them.

If this analogy holds, then the elusive "Grand Unified Theory" physicists have been seeking is not a single equation describing reality's ultimate substance but a description of the **architecture of the renderer itself**. What we think of as "laws of nature" would then be high-level abstractions — callable functions in the system's code — optimized for stability, efficiency, and coherence rather than for satisfying human notions of elegance.

This is the shift Part I aims to solidify. It asks the reader to see not a scattered set of puzzles but a single, integrated model — a **unified rendering model** — in which physics' deepest anomalies are no longer bugs but documented features of the engine running the cosmos.

Chapter 3
Why Gravity Is Different

Gravity as Emergent Geometry in a Computational Fabric

Among the four fundamental interactions, gravity stands apart like a law written in a different language. Electromagnetism, the weak force, and the strong force all yield—at least in principle—to the grammar of quantum theory. They can be described in terms of field quanta and gauge symmetries, their effects traced through discrete exchanges of energy packets across an underlying spacetime grid. They live *within* the world.

Gravity, however, behaves as though it *is* the world. When general relativity replaced Newton's notion of a force with that of spacetime curvature, Einstein effectively declared that gravity was not a messenger but the medium itself. Every other interaction presumes a preexisting stage; gravity *defines* the stage. That fundamental distinction has haunted theoretical physics ever since. The quantum field formalism assumes a background on which probabilities can play out. Gravity, by definition, is the background. To quantize it would be to make the canvas itself shimmer with uncertainty — a contradiction that has yet to resolve.

In a simulation framework, this resistance is not an obstacle but a clue. Gravity's refusal to quantize hints that it may not belong to the same category of processes as the other forces. If electromagnetism and the nuclear forces are subroutines — modular systems handling interactions, bindings, and decays within a rendered environment — then gravity is not another subroutine. It is the rendering engine's *global geometry loop*.

Gravity, in this view, is the computational process by which spatial continuity and scale coherence are maintained. It is not a discrete signal but a continuous calculation, updating the relationships between objects to preserve a consistent geometry across frames of simulation. This is why it appears universal: every entity in the simulation participates in the same geometric computation. The "mass" of an object, then, is not merely an intrinsic property but a measure of how much that object contributes to the computational load of maintaining local curvature. The more mass-energy concentrated in a region, the more the geometry engine must adjust its metrics to preserve coherence — an effect we interpret as gravitational attraction.

From this standpoint, relativity's spacetime curvature and Newton's inverse-square force are two descriptions of the same underlying operation: one mathematical, one emergent. What we call "gravitational force" is simply the visible artifact of the geometry engine maintaining equilibrium within a dynamic lattice of rendered space. It does not pull or push — it recalculates.

This interpretation also reframes gravity's weakness relative to the other forces. In quantum field terms, it is anomalously feeble; a single charged particle's electromagnetic field overwhelms its gravitational influence by some forty orders of magnitude. But if gravity is not a force propagated through quanta, its weakness is not a flaw — it is a feature of its scope. The geometry routine must operate at the most efficient computational scale possible; it is distributed across the entire simulation grid. Each local instance contributes infinitesimally to the global geometry, producing an effect that is both omnipresent and extremely subtle at the local level.

In this sense, gravity behaves not as a fundamental interaction but as an emergent *consistency condition.* It enforces the rules that make all other interactions possible, ensuring that distances remain measurable, causality remains coherent, and the fabric of space retains a continuous topology even as it flexes under energy and mass. To "quantize gravity," therefore, may be akin to trying to identify the discrete units of a rendering engine's coordinate system — a meaningful mathematical exercise, perhaps, but one that mistakes implementation detail for ontology.

If the universe is a rendering, gravity is not the play of forces within it — it is the algorithm that draws the grid, smooths the edges, and keeps the stage coherent as the simulation evolves. It is less a participant in the cosmic drama than the rule by which the performance takes form.

Is Spacetime "Stored" or Dynamically Calculated?

Here lies one of the most revealing questions in the simulation interpretation of physics: is spacetime a **pre-rendered archive** — a vast, immutable database that exists in full whether or not it is accessed — or is it a **procedurally generated construct**, calculated dynamically and locally as interactions require? The distinction is not academic. It cuts to the heart of whether the universe behaves more like a static storage medium or an active computation.

The Stored Model: A Universe Written to Disk

In the stored model, spacetime is akin to a pre-rendered map — a world "written to disk" at initialization. Every coordinate, every curvature, every quantum fluctuation would exist in detail regardless of whether any observer ever encounters it. The universe would therefore be a total dataset, complete in its specification and fixed in its content. Gravity, in such a schema, could be understood as a **static field**: a precomputed gradient distributed across this cosmic memory.

This picture has a certain intuitive appeal because it mirrors the classical worldview. Newton's universe functioned like clockwork — every motion, every future event implicitly contained within initial conditions. Even Einstein's spacetime, though dynamic in form, still presupposes a complete four-dimensional manifold: past, present, and future all coexisting as coordinates on a single geometrical block. To "experience" time, in this framework, is merely to move along a pre-existing dimension.

But the stored model bears a steep computational cost. To maintain the full state of the universe — every particle, every fluctuation, every unobserved star and unmeasured vacuum region — would require a storage capacity beyond comprehension. Even with hypothetical compression or emergent information principles, a fully pre-rendered cosmos would entail inefficiency on a cosmic scale.

It also conflicts with one of the central discoveries of the quantum age: **unobserved systems do not manifest in definite states**. The wavefunction describes probability amplitudes, not certainties. Only when an observation occurs does the system "collapse" into a definite outcome. In a stored model, collapse would be unnecessary — the result would already exist. That violates both the spirit and the mathematics of quantum mechanics.

The Dynamic Model: A Universe Rendered in Real Time

The alternative — and arguably more elegant — view is that spacetime is not stored, but **computed in real time**. In this dynamic model, the universe behaves less like a static archive and more like a live rendering engine. The "fabric" of reality is not pre-laid; it is generated *on demand*, where and when interaction requires it.

Under this paradigm, observation and interaction become the triggers that call the rendering subroutines. A photon's path through space is not mapped out in advance; it is calculated as it interacts with the geometry defined by nearby energy distributions. The same logic extends to the large-scale structure of the cosmos: the curvature of spacetime is not written into the codebase but **solved continuously** as the simulation evolves.

This approach mirrors relativity's insight that spacetime is not a fixed container but a relational structure defined by mass-energy distributions. Yet it extends the principle further: the structure is not merely flexible but ephemeral, existing only insofar as it is needed to maintain coherence between events. When no observation or causal chain connects two regions, there is no need for the engine to maintain their geometry in memory. They exist only as **potential coordinates**, not instantiated data.

The computational implications are profound. The dynamic model minimizes storage requirements by leveraging **procedural generation** — the same principle that allows modern game engines to render vast worlds from compact rule sets. The laws of physics serve as compression algorithms: instead of storing the universe, the engine stores only the *rules* for generating it. Reality becomes a self-updating solution to the ongoing problem of maintaining internal consistency.

Runtime Geometry and the Nature of Gravity

Within this dynamic framework, gravity ceases to be a static curvature frozen into a cosmic manifold. Instead, it becomes a **runtime geometry solver** — a process continuously recalculating the optimal configuration of spacetime to maintain continuity across frames of computation. Mass and energy are not objects embedded in a background; they are *inputs* to this solver.

Every tick of the simulation clock triggers a new iteration of the geometry solution, updating curvatures according to energy densities and the demands of causality. In this sense, gravity is the act of *rendering continuity*. It ensures that local interactions cohere into a global structure that appears smooth, lawful, and continuous — even though the underlying process may be discrete and algorithmic.

This model offers a natural reconciliation between quantum indeterminacy and relativistic continuity. Quantum events determine *where* the rendering engine must resolve detail; relativity determines *how* that detail must be geometrically consistent across the whole. Together they describe a universe that is neither stored nor static, but continuously computed — a simulation that optimizes its own existence moment by moment.

Equivalence Principle as a Shortcut for Rendering Inertia

Einstein's *equivalence principle* is one of the most elegant and far-reaching insights in the history of physics. It states, simply, that the effects of gravity and the effects of acceleration are indistinguishable to any local observer. A person sealed in a windowless elevator cannot tell whether the force they feel pressing them to the floor is due to gravity or due to the elevator accelerating upward. In Einstein's hands, this principle became the cornerstone of general relativity — the key insight that replaced Newton's "force of gravity" with spacetime curvature.

Yet viewed through a computational lens, the equivalence principle takes on an additional — and intriguingly pragmatic — significance. What Einstein discovered as a deep truth about the structure of nature may also reflect something like an *optimization routine* in the universe's code.

The Programmer's Shortcut

In software, redundancy is costly. Maintaining two distinct subroutines to handle similar phenomena wastes processing power and increases the risk of inconsistency. A well-designed engine will consolidate overlapping functions into a single, more general instruction set.

From this perspective, gravity and inertia are not two distinct processes but two expressions of the same underlying computational rule — a unified instruction governing how matter interacts with curvature in the spacetime grid. Instead of calculating one set of transformations for "free fall" and another for "acceleration," the system uses one algorithm that produces identical outputs in both contexts.

In effect, the equivalence principle becomes a **shortcut for rendering motion**. It tells the engine: "If an observer experiences acceleration, treat it as a gravitational effect; if an observer is in a gravitational field, treat it as acceleration." The distinction collapses — not because of metaphysical unity, but because of *computational efficiency*.

The Elegance of Economy

Einstein often remarked on the aesthetic beauty of physical law — its simplicity, coherence, and economy. In a simulation framework, that elegance is precisely what one would expect from an optimized system architecture. The universe, if rendered, would favor the least redundant, most parsimonious implementation possible. Every symmetry, every invariance, every apparent "law" may be a clue to the engine's commitment to efficiency.

Under this view, the equivalence principle is not just an empirical fact; it is an **architectural feature**. The simulation doesn't separately "calculate" gravitational force; it computes geodesic motion — the natural path through a dynamically curved geometry — using one unified rendering rule. Whether you're in orbit or in an accelerating rocket, the visual and physical outputs are identical because the same subroutine handles both cases.

Gravity as Geometry, Inertia as Default Motion

General relativity tells us that gravity is not a force at all, but the manifestation of motion along curved spacetime. The equivalence principle makes this clear: an object in free fall isn't "pulled" by gravity; it's simply following the straightest possible path through a curved environment. In a rendered universe, this could mean that inertia — the tendency of an object to maintain its motion — is not a separate phenomenon, but the **baseline behavior of the geometry engine** when no additional instructions are applied.

Gravity then becomes a change in the parameters of that geometry, a real-time adjustment of the rendering matrix that modifies what "straight-line motion" looks like locally. The system doesn't need to apply a new force; it only needs to slightly reshape the space through which the object moves.

A Window into the Code

Seen this way, the equivalence principle may represent more than a unification of physical experience — it may be a window into the logic of the underlying code. It is the universe revealing its hand, showing that at some deep level, motion, curvature, and acceleration are computed by the same subroutine.

This would explain why general relativity has such mathematical grace. The equations of motion derived from Einstein's field equations are astonishingly compact — they describe trajectories, curvature, and acceleration with minimal redundancy. That efficiency, so long admired as a hallmark of physical truth, might instead be a sign of **elegant programming**.

From Physics to Architecture

If the equivalence principle reflects a computational shortcut, then general relativity's success is not accidental. It is the natural consequence of how the rendering engine optimizes the display of motion and geometry. The universe's "laws" are not necessarily ultimate truths about existence but the **best-performing algorithms** for maintaining coherence in a finite, resource-constrained system.

Thus, the equivalence principle is not merely a profound statement about physics; it is also an invitation to think like an engineer. Why does the universe behave so economically? Why are its rules so elegantly minimal? Perhaps because, at the most fundamental level, *efficiency is the law.*

Interlude

What the Sim Would Look Like From the Inside

Let us pause the technical argument for a moment and step into a more imaginative space. The preceding chapters have spoken in the language of physics and computation — of curvature and clock rates, Planck pixels and runtime constraints. But theory alone can become abstract to the point of disconnection. To grasp what a rendered reality might *feel* like from within, we must momentarily inhabit it.

So imagine, not as a distant observer but as a participant, that you live inside such a system — that you are a conscious process running on the substrate of some vast, unseen computation. You cannot exit the world to inspect its source code, but you can infer its nature by what you notice — by the seams, the optimizations, the way detail appears only when demanded by attention. This interlude is a thought experiment: a glimpse of what the simulation might *look like from the inside.*

The Scene

You stand at the edge of a cliff, overlooking the sea. The air is cool and fragrant with salt. The horizon stretches out like a promise — seamless, endless, and yet curiously finite. You sense, though you cannot prove, that beyond that line the world continues only as potential, waiting to be called into being. The sky is a smooth gradient, unblemished by imperfection, as if painted by an algorithm whose purpose is to suggest infinity, not to model it.

Below, the waves roll and shatter against the rocks. Foam blossoms, spray catches the light. Yet you know, instinctively, that these details exist only because you are looking. If you close your eyes, the sound remains — but the motion itself, the exact shape of each wave, dissolves. The sea becomes a remembered field, a set of parameters ready to re-render when next you turn your gaze.

You walk inland, toward the forest. At the treeline, the world feels tangible: bark rough against your hand, moss yielding beneath your feet. Light filters through the canopy in precise shafts, the shadows crisp and definite. But as you venture deeper, something strange occurs. The edges soften. The distant trees blur into abstraction, more texture than object. The forest seems to conserve its complexity, rendering only what your senses require.

A bird darts across your field of vision — vibrant, momentary. You track its motion as it disappears into the green haze above. You could swear that, once it leaves your sight, it ceases to be *resolved*. Perhaps it exists only as a trajectory stored in memory, a variable marked "in flight." Were you to look again, it would reappear, the details freshly computed, consistent with everything you remember.

You feel, faintly, the weight of the world as computation — not cold or mechanical, but curiously elegant. The universe gives you exactly as much as you need to experience it: no more, no less. The world's richness is not a measure of stored data but of responsiveness. Reality is dynamic, contingent, called into existence by curiosity itself.

Resolution and Attention

In such a world, *attention* becomes the true engine of existence. The act of perceiving is an act of rendering; the act of remembering, a way of caching. You begin to notice how the universe economizes. Distant objects shimmer with less definition. Sound fades before detail does. Even your thoughts — sharp when you focus, diffuse when you drift — seem to follow the same principle. The system optimizes for relevance.

And then you wonder: if this were truly a simulation, would anything feel different? The air is still cool on your face. The forest still hums with unseen life. Whether computed or continuous, the world still feels real because *you are part of its computation*. The simulation does not exist "out there." It is a process that includes you — your perception, your memory, your sense of time.

The horizon, then, is not a limit of distance, but of awareness. The seams in the code are not visible in the sky, but in the edges of comprehension — in the places where certainty blurs into probability, where detail dissolves into potential.

An Invisible Boundary

If you were to stand there long enough, watching the waves and the wind, you might begin to notice that the world feels less like a place and more like a conversation. You ask a question with your senses, and the universe answers with form. You look for structure, and structure appears. You stop looking, and it recedes.

And perhaps that is the most profound revelation of all: that the universe's apparent objectivity — its solidity, its independence — may be a user interface. A shared rendering space, dynamically negotiated between countless observers, each sampling from the same underlying code.

From within, it feels natural — inevitable, even. Only when one learns to see the faint shimmer at the edge of experience does the suspicion arise: that the world is not *given*, but *generated.*

The Realization

At first, you dismiss these impressions as tricks of perception. Surely your eyes play games, your mind fills gaps, your brain filters what it cannot process in full. That is, after all, what perception is—a compression algorithm optimized by evolution. But then you begin to notice the patterns.

The way light bends predictably near heavy stones, always following the same invisible gradient, as though obeying not a natural inevitability but a prewritten function. The way your shadow elongates in perfect ratio as the sun lowers, never once breaking the mathematical rule, no matter the angle or the hour. And the way distant stars flicker with suspicious regularity—each twinkle like a single pixel trying to approximate infinity at the edge of a great dome.

You begin to feel it not as a thought but as a tension in the air: reality itself is running smoothly. Too smoothly. The chaos that should accompany a living universe—the small errors, the messy unpredictabilities—seems strangely absent. Instead, everything flows within the constraints of elegant, recursive simplicity. The more you look, the more the world resembles not a spontaneous cosmos but a simulation running on exquisitely efficient code.

The laws you have been taught to trust—gravity, thermodynamics, electromagnetism—begin to look less like cosmic commandments and more like optimization routines. They conserve energy not out of moral necessity, but out of computational prudence. They are not the handwriting of gods; they are the parameters of a system trying to maintain coherence.

At times, the realization feels liberating. The veil lifts, and what had been mystery now becomes pattern. You start to see the logic behind the beauty: how every gust of wind might be a calculation resolving fluid dynamics in real time, how every reflection in water is a fractal recursion optimized for perceptual believability. You marvel at the sophistication of the rendering, the seamless integration of perception and expectation.

But then the other feeling creeps in—the quiet dread. Because if the world is an algorithm, where are *you* in it? Are you an observer or an instance? A consciousness or a process? If even your thoughts arise from code, then what does it mean to notice the code at all? Is that awareness itself an emergent function—something the system allows, or an unplanned glitch?

You look again at the horizon, at the smooth curve of the sea meeting the sky, and the sensation becomes almost unbearable: the suspicion that everything you see, touch, and feel exists only when observed. That existence itself may be a rendering trick, a form of deferred computation—real enough for experience, but never more than what needs to be.

And in that moment, standing between belief and simulation, you no longer know whether you've glimpsed the truth or merely stared too long into the structure of your own perception.

The Phenomenology of Rendered Existence

The exercise of imagination just performed was not mere indulgence. It was, in miniature, a phenomenological experiment: a way of describing *how* a rendered universe would feel from the inside, given the perceptual and physical structures already known to govern our experience. What emerges is an uncanny resonance between the imagined and the actual—the subjective texture of reality as we experience it and the objective behavior of the physical laws that structure it.

When we look closely, three correspondences stand out.

Quantum Observation and Deferred Detail

In the imagined scene, reality "loads" only where attention rests—the ocean fades when turned away from, the forest blurs into abstraction until approached. This parallels the core of quantum mechanics: the observer effect, where physical systems do not take on definite values until measured. The so-called *collapse of the wavefunction* may not be collapse at all but computation—an on-demand rendering of the world's state when queried by observation.

Just as a graphics engine does not waste resources on off-screen geometry, a reality-processor might not instantiate definite positions or energies for particles until interaction requires it. Quantum indeterminacy, from this angle, is not mystery but optimization.

Cognitive Limitation as Bandwidth Management

Human perception is bounded by thresholds—of light, color, sound, distance, and time. We never perceive *everything* in the frame; our brains compress, interpolate, and discard vast amounts of data. In a rendered universe, this would not be an accident of biology but an extension of the same efficiency principle. The perceptual system and the rendering system would be co-optimized.

Neuroscience already tells us the brain functions as a *predictive modeler*: it renders probable futures and fills in missing information, using attention as a focusing algorithm. In phenomenological terms, consciousness is itself a renderer—one nested within another. The limits of cognition and the limits of physical resolution may, therefore, be reflections of the same architectural choice: to privilege coherence and experience over totality and precision.

Relativity and the Geometry of Experience

The equivalence between gravity and acceleration—Einstein's most elegant insight—also reads differently through this lens. If spacetime is not a fixed grid but a runtime geometry, then relative motion is not an illusion of perspective; it *is* the recalculation of geometry in response to differing vantage points.

Time dilation and length contraction, which seem so alien to intuition, could be seen as computational trade-offs: the cost of maintaining consistency across locally rendered reference frames. What we perceive as curved spacetime may simply be the visual artifact of synchronizing distributed computation into a single, self-consistent simulation.

Together, these correspondences form a phenomenology of rendered existence—a description of how a computational cosmos would necessarily *appear* to its inhabitants. The subtle flicker of quantum randomness, the compression of sensory reality, the elasticity of spacetime—all of these could emerge as signatures of a system optimized for stability and coherence from the inside.

This is not an argument that we live in such a system. It is a recognition that our experience of "reality" already behaves *as if* it were rendered. Phenomenology, cognition, and physics converge on the same design logic: instantiate what is needed, when it is needed, to preserve continuity of experience.

As we move into the next chapter, we turn from description to explanation—from what rendered reality *feels like* to what its *architecture* might require. If the simulation analogy holds, then the task ahead is not to prove the universe "is" a simulation, but to identify the computational principles that could make such a universe indistinguishable from what we already call the real.

The Implication

If the universe is rendered, then reality is not a fixed tableau but a process — an ongoing act of computation that unfolds only as required. This carries profound implications for how we understand physics, consciousness, and causality.

First, it means that **laws are not prescriptions but algorithms**. The uniformity of nature — that gravity behaves identically on Earth and in a distant galaxy — reflects not metaphysical necessity, but code consistency. A good simulation must run identically across all instances; the universality of physics may be an artifact of that requirement.

Second, **matter and energy would be emergent, not fundamental**. They would exist as the "compiled output" of deeper informational structures. What we perceive as particles and fields could be the projections of data structures optimized for stability and interaction. This reframes the search for a Theory of Everything: the unifying principle may not be a force or field, but the architecture of computation itself.

Third, **observation and consciousness take on central roles**. In a rendered system, observation isn't passive — it is the trigger that calls the code. Every act of perception is a request to the rendering engine to resolve uncertainty into experience. Quantum indeterminacy and the observer effect would thus not be quirks of measurement, but expressions of the system's efficiency: reality manifests only when queried.

Finally, **causality becomes layered rather than linear**. What we experience as temporal progression might be the emergent order of processed frames, while the underlying system operates outside that timeline — perhaps recalculating or iterating solutions in ways that our sequential perception cannot detect. This hints that certain "cosmic coincidences" or finely tuned constants might not be coincidences at all, but deliberate optimizations ensuring the simulation remains stable and intelligible to its inhabitants.

In sum, the implication is this: if we live in a simulation, then physics is not a description of ultimate reality — it is documentation of the rendering system. Every constant, every symmetry, every conserved quantity is not evidence of necessity but of design. The goal of science, then, shifts from discovering immutable truths to reverse-engineering the logic of the universe itself.

The Test

If such a framework holds even partially true, then it should leave fingerprints — not proof in the metaphysical sense, but anomalies that point toward an underlying computational substrate. How, then, could one test the hypothesis?

Resolution Limits and Discreteness.
If spacetime is quantized at the Planck scale, then we might detect deviations from smooth continuity in high-energy regimes. Searches for "lattice artifacts" — irregularities in the behavior of cosmic rays, or anisotropies in spacetime at extreme energies — could reveal the granularity of the grid. If the universe has a finite resolution, it may also exhibit frame-rate artifacts, such as

unexplained cutoffs in particle behavior beyond certain thresholds.

Information Conservation and Error Correction. James Gates' discovery of error-correcting codes within supersymmetry equations hints at something extraordinary: code-like redundancy embedded in the mathematics of nature. If physical law encodes built-in correction protocols — ensuring that information is never truly lost, even across black holes — this could be evidence of a system maintaining internal coherence, much like digital error-checking in computation.

Quantum Randomness and Pseudorandomness. Quantum phenomena appear random, but in a simulated environment, true randomness is difficult to achieve. Instead, we might find statistical signatures of pseudorandomness — patterns that appear random but are generated algorithmically. Subtle correlations in supposedly uncorrelated events could hint at a shared computational source.

Fine-Tuning and Optimization. The remarkable precision of fundamental constants — the narrow margins that allow matter, chemistry, and life to exist — may be indicative of optimization rather than accident. If the universe's parameters are tuned as if by iterative refinement, the fine-tuning problem ceases to be a mystery of chance and becomes evidence of system calibration.

Computational Constraints and Cosmic Energy Budgets. If the universe runs on finite resources, there may be detectable signatures of computational load balancing. The constancy of the speed of light, the cosmic speed limit, might represent the maximum information throughput of the system. Likewise, the expansion rate of the universe — its accelerating growth — could represent dynamic reallocation of computational resources as entropy increases.

Bridging Thought and Experiment

If we are characters inside a rendering, we cannot "see" the processor itself — but we can infer its presence through the regularities and constraints it imposes. The task, then, is not to break the illusion but to **read it**: to find the seams in the fabric and decode their logic.

Just as early physicists inferred the existence of atoms, or cosmologists inferred dark matter from gravitational behavior, so too might we infer the substrate of reality through its computational signatures. The test of the Simulation Hypothesis is not whether we can escape it, but whether we can recognize its fingerprints within the physics we already know.

Returning to the Argument

This imaginative vignette is not meant as proof, nor even as a direct claim about reality. It is, instead, a kind of *conceptual breather*—a pause in which we step out of technical language and step into the skin of the hypothesis itself. By momentarily inhabiting the Simulation Hypothesis from within, we gain not evidence but perspective: a sense of how such a universe would feel to its inhabitants, how its seams might remain invisible until examined with care, and how the ordinary might suddenly appear algorithmic under a shift of attention.

If the universe were a rendered system, this is precisely the texture it would offer—an environment seamless at human scales, but economical at the margins; richly detailed in the foreground, but suspiciously vague at the edges. It would not announce itself as an artificial construct. It would present itself as "the real," because that is precisely what an optimized system must do for its participants: deliver a coherent, lived experience rather than a transparent map of its underlying machinery.

This is why the thought experiment matters. Not because it proves anything, but because it reframes the way we interpret the data we already have. When you imagine the world as a rendering pipeline, every smooth regularity, every "law of nature," every apparent limit on speed, scale, or resolution becomes not merely a constraint but potentially a clue. It invites us to re-read the familiar not as brute necessity but as design choice.

As we now turn to Part II, we leave imagination behind and return to analysis. We will resume the tools of argument—physics, computation, and epistemology—to evaluate the Simulation Hypothesis on its actual merits. But the image lingers. The world as a rendered landscape, optimized for our perception. The edges of reality as deferred computation, waiting to resolve until called upon.

Whether or not one accepts the hypothesis, that mental picture now sits in the background, quietly altering the way evidence is weighed. And that is the point of the interlude: to plant a lens through which the coming analysis may be viewed, not as an abstract proposition but as a lived possibility.

Part II

Rethinking GUT in a Simulated Framework

For over a century, the search for a **Grand Unified Theory (GUT)** has been one of science's most ambitious and alluring quests — the attempt to find a single framework that unites all known forces and phenomena under one coherent mathematical law. From Einstein's dream of a unified field to the elegant symmetries of string theory, physicists have pursued the conviction that reality, at its deepest level, must be one.

This pursuit has assumed a shared premise: that the universe is a **continuous, physical fabric**, a single ontological layer governed by universal principles. The task, then, has been to reconcile two great pillars — quantum mechanics and general relativity — into one seamless description.

Yet after decades of ingenuity, the unification remains tantalizingly out of reach. The quantum realm and the relativistic cosmos remain divided not by ignorance, but by **incompatibility**.

- Quantum mechanics operates in a probabilistic domain, built from discrete quanta of energy and information.
- General relativity describes a deterministic continuum, where space and time curve smoothly in response to mass and energy.

One deals in uncertainty and wavefunctions; the other in geometry and curvature. Attempts to reconcile them — whether through string theory's multidimensional harmonies, loop quantum gravity's discretized spacetime, or other exotic proposals — have yielded exquisite mathematics but little experimental traction. The deeper we dig, the more the two frameworks seem to speak incompatible dialects.

Perhaps, then, this enduring schism is not a **temporary technical challenge**, but a **category error**. What if the obstacle lies not in our mathematics, but in our ontology?

Suppose quantum mechanics and relativity do not describe two aspects of a single continuous reality — but rather **two distinct computational layers** within a simulated architecture. In such a model, the failure of unification would not signify an incomplete theory, but rather a **design feature** of the system itself. The micro and the macro would not be forced to obey one equation because they are not running on the same codebase.

In this light, the apparent incompatibility between quantum and relativistic physics becomes diagnostic — a signature of layered computation. Quantum processes might represent **low-level rendering routines**, the pixel-level updates that generate localized phenomena. Relativistic geometry, by contrast, could be the **macro-engine** — a higher-order process governing how the rendered world maintains continuity and coherence across vast scales. Their discord is not a flaw to be resolved but a clue to their distinct computational functions.

Thus, in this Part, we step beyond the traditional question — *"What equation unites all forces?"* — and ask instead:
"What architecture coordinates their operation?"

By reframing the Grand Unified Theory as a question of system design rather than physical law, we open a new investigative frontier. The unification we seek may not be a formula inscribed in nature, but an emergent order arising from the logic of a cosmic rendering system.

If so, the divide between quantum and relativistic domains is not a failure of physics — it is the **map of the engine itself**.

Chapter 4
The Failure of Unification
A Feature, Not a Bug

The Mismatch Between Probabilistic and Deterministic Formalisms

At the very center of modern physics lies a deep and persistent fracture — one that has resisted every attempt at reconciliation for nearly a century. It is not a disagreement about data, but about *ontology* — about what kind of thing reality fundamentally is.

On one side stands **quantum mechanics**, the mathematics of uncertainty. It describes a world where entities do not exist in definite states until measured, where probabilities replace certainties, and where reality itself seems to shimmer in and out of definition. Its laws are not violated, but they are *statistical*: the outcomes of countless micro-events can be predicted, but individual occurrences remain irreducibly indeterminate.

On the other side stands **general relativity**, the mathematics of inevitability. It describes a universe of elegant continuity — spacetime as a smooth, four-dimensional manifold in which matter and energy dictate curvature, and curvature dictates motion. Nothing in relativity is probabilistic. It is a theory of graceful determinism, where every event follows necessarily from geometry.

Between these two visions lies a profound incompatibility:

- **Quantum mechanics** assumes that nature at its core is *discrete, discontinuous,* and probabilistic.

- **Relativity** assumes that nature at its core is *continuous, differentiable,* and deterministic.

Their mathematical languages do not speak to each other. Quantum mechanics operates in Hilbert space, with complex probability amplitudes that collapse under measurement; relativity operates in spacetime, with tensors that curve in response to mass-energy. Efforts to merge them — from quantum field theory's renormalizations to string theory's higher dimensions — often look less like synthesis and more like patchwork. One theory demands granularity; the other demands smoothness.

It is as if two incompatible engines are running side by side, producing a single coherent experience — but never truly merging into one algorithm.

From a **simulation framework**, this is not only understandable but expected. It suggests that what physicists call the "quantum world" and the "relativistic world" are in fact **two computational domains**, each optimized for a distinct level of simulation fidelity.

The **quantum domain** functions as a **probabilistic rendering engine** — a module designed for local efficiency. It does not calculate every particle trajectory in advance; it calculates probabilities, then resolves specifics only when necessary (i.e., upon observation or interaction). This minimizes computational overhead, conserving processing power for events that actually affect the observer's frame.

The **relativistic domain**, by contrast, functions as a **deterministic geometry engine** — a module designed for global consistency. It ensures that space, time, and causality remain coherent across the entire simulation. Its equations maintain the smooth continuity of the "world stage," even when the underlying pixels — the quantum events — flicker in and out of computation.

Seen through this lens, the incompatibility between quantum mechanics and general relativity is not a paradox of nature but a **division of computational labor**.

Each system obeys its own internal logic, optimized for its role in maintaining a stable, rendered universe. Their disagreement is not a sign of error but a reflection of **modular design** — like different subsystems in a complex program, each handling specialized tasks according to its own constraints.

The physicist's quest for unification — the attempt to compress these domains into a single formalism — might therefore be akin to forcing two software libraries written in different programming languages into one file. It can be done in principle, but the result is not simplification; it is convolution. The real insight comes not from fusing them, but from **understanding the architecture that coordinates them**.

From this perspective, the "failure" of unification is one of the most revealing signatures of all. It tells us that we are not dealing with one seamless physical substrate, but with **a layered computational hierarchy** — one probabilistic and localized, the other deterministic and global. The universe, in other words, may be built not from one set of laws but from *two cooperating systems*, unified not by equation, but by code.

Why Gravity Resists Quantization in a Rendered Environment

If gravity is not a force but the geometry engine itself (as argued in Chapter 3), then the failure to quantize it is not an embarrassment of physics — it is a diagnostic clue. Quantization presumes that a phenomenon arises from discrete, local interactions that can be counted, exchanged, and measured. It presumes that something happens *within* the field of play. Gravity, by contrast, is the curvature of that field itself — not a piece within it, but the rulebook that defines what counts as "within."

To attempt to quantize gravity, then, is to attempt to treat the game engine as if it were a sprite. It is to apply the mathematics of emergent phenomena to the substrate that makes emergence possible in the first place. This is not simply the wrong equation; it is the wrong category of equation.

In a rendered environment, quantization describes *what happens when the engine draws detail*. Each pixel or particle represents a discrete computational event — an update to the probability space as it collapses into a definite outcome. Gravity, on the other hand, is not one of these events. It is the logic governing *how space itself is drawn* — how relationships between rendered points maintain coherence as new data are generated. To quantize gravity would be to insist that the *canvas* behave like the *paint*.

This also reframes why gravity interacts so weakly compared to the other fundamental forces. In a simulated architecture, the rendering engine is not "another subsystem" running alongside electromagnetism or the strong and weak nuclear forces. It is the environment in which those forces operate. Any attempt to detect a graviton — a "quantum of spacetime curvature" — is therefore analogous to looking for a pixel that encodes the monitor itself. No matter how fine the resolution of our instruments, we remain inside the display. The substrate cannot be directly rendered within its own rendering.

From this perspective, the "failure" of quantum gravity research is not a failure of ingenuity but a failure of scope. It is a misunderstanding of what gravity *is* in relation to the rest of physics. The deeper implication is that quantization and geometry may be complementary expressions of different computational strategies:

- **Quantization** handles the *local economy* of events — the pixel-level calculations where energy and probability are balanced on the fly.
- **Geometry** handles the *global consistency* of the world — the constraints that ensure those pixel updates still conform to a coherent, continuous spacetime.

In this view, gravity resists quantization for the same reason that the rendering pipeline in a video game cannot be meaningfully decomposed into particle interactions. It is not built from them — it *builds* them. To seek a "graviton" is to look for a brick in the foundation that is, by design, not made of bricks at all.

Thus, the enduring mismatch between general relativity and quantum mechanics ceases to be a riddle. It becomes evidence of a layered architecture: probabilistic microstates computed locally, deterministic continuity maintained globally. The boundary between them — where quantum indeterminacy appears to "average out" into smooth spacetime — is not a flaw in our mathematics. It is precisely what we should expect from a system optimized to render only as much reality as needed to sustain the illusion of continuity.

"Local Rules, Global Coherence": How a Simulation Separates Levels of Reality

A well-designed simulation must satisfy two seemingly contradictory imperatives. On the one hand, it must handle an immense number of micro-level interactions without overloading its computational capacity. On the other hand, it must ensure that the world it produces appears seamless, lawful, and stable to any observer within it. This tension — between *local rules* and *global coherence* — is not an accident but a design strategy.

Local Rules
Efficient computation favors locality. At the smallest scales, the system runs probabilistic routines, updating only what is necessary. Particles are not fully rendered objects but probability distributions, potentialities waiting to be "collapsed" into definite states when conditions demand it — when, for example, an observation occurs or an interaction makes further specification unavoidable. This is the micro-level's economy of calculation: by working with probabilities instead of concrete values, the system conserves processing power. It does not calculate every possible path; it calculates only the branch that becomes actual.

Global Coherence
Yet this frugality must not compromise the larger illusion. To the inhabitants of the simulation, the macro-world must appear stable and predictable. Galaxies must hold together. Orbits must close. Energy and momentum must be conserved. Deterministic geometry routines provide this consistency. They act as the scaffolding on which probabilistic events unfold, ensuring that when local outcomes are rendered, they still fit into a coherent

whole. The "canvas" stays taut even as the "paint" is applied pixel by pixel.

This separation of levels — probabilistic micro, deterministic macro — is not a defect waiting to be solved by a unifying equation. It is an architectural choice. By dividing reality into layers with different computational logics, the system gains speed, stability, and flexibility. Each layer is optimized for a different task: the micro-level for efficiency and indeterminacy, the macro-level for continuity and predictability. Their mismatch is not a bug but a feature.

From this perspective, the ongoing failure of unification in physics becomes diagnostic. It signals that we are not dealing with one continuous mathematical fabric but with a layered computational construct. Quantum mechanics and general relativity do not disagree because one of them is wrong; they disagree because they describe different modules of the same deeper architecture. The absence of a single unifying equation is itself a kind of evidence — not of our ignorance, but of the system's design. It is exactly what we should expect if reality is rendered on demand, with one engine handling the pixel-level probabilities and another maintaining the integrity of the stage on which those pixels appear.

Chapter 5
Physical Laws as Code Abstractions

Why the Macro World Is Smooth and the Micro Is Granular

In a simulated framework, the striking difference between the *smoothness* of the macroscopic world and the *granularity* of the microscopic is not a mystery of physics but an artifact of optimization. What appears to us as two incompatible regimes — continuous versus discrete, deterministic versus probabilistic — may instead reflect two different coding strategies serving two different performance goals within the same architecture.

The Macro World: Smooth by Design

At human scales, the world appears continuous. Tables don't flicker between probabilities; oceans don't phase in and out of existence when no one looks. The orbits of planets trace reliable ellipses, and the curvature of spacetime follows the elegant continuity of general relativity.

In a simulation framework, this smoothness is not because the system calculates every atom's position in real time. That would be computationally impossible. Rather, macroscopic objects are rendered as *aggregates* — large-scale data structures or "meshified" constructs that behave as if continuous. Much like a mountain in a video game, which can be displayed as a single mesh rather than millions of individual rocks, macro objects are represented through efficient abstractions. Their apparent continuity is an illusion created by scalable resolution: when you zoom in, more detail can be generated procedurally, but when you step back, the system renders only what's necessary to maintain coherence.

This approach economizes processing power while preserving experiential realism. From the perspective of an observer within the simulation, the macro world *must* be smooth; otherwise, reality would appear jagged, inconsistent, or unstable. Continuity is not a property of matter—it is a function of rendering stability.

The Micro World: Granular by Necessity
At smaller scales, however, the system faces a different constraint. There are far too many potential microstates to precompute them all. Instead, it adopts a probabilistic routine: quantum mechanics. Here, "granularity" — the discrete quanta of energy, charge, or spin — emerges as the signature of a deferred computation. The simulation does not specify each state until required; it holds the possible configurations as probability amplitudes, awaiting the trigger of observation or interaction to resolve them into definite outcomes.

This mechanism mirrors the logic of modern rendering systems. A graphics engine does not compute every pixel of an unseen landscape; it waits until the player turns the camera. The act of observation is, in effect, the *render call*. Before that call, the system tracks potentialities rather than actualities — the quantum wavefunction instead of the particle. Collapse, then, is the act of finalizing the render at the point of need.

The Optimization Principle
From this perspective, the divide between the smooth macro and the granular micro is not a contradiction but a hierarchy of abstraction. The macroscopic world is optimized for perceptual stability; the microscopic world is optimized for computational

efficiency. One ensures the continuity of experience, the other the conservation of processing power.

If physical law is code, then what we call "quantization" and "continuity" are not competing truths but complementary programming paradigms — discrete and analog — interleaved for efficiency. Each does what the other cannot: the micro ensures adaptability and resource management; the macro ensures coherence and intelligibility.

The tension between them is not a flaw in nature's design but a clue to its architecture. It suggests that what we perceive as physical laws are, in fact, code abstractions — emergent regularities produced by a higher-order logic that balances fidelity with computational cost.

Physics as API: Repeatable, Callable Behavior in Response to Observation

Every software system relies on *interfaces*—well-defined points of contact between user and code. In programming, these are APIs: **Application Programming Interfaces**. An API specifies what a function does, the parameters it takes, and what it will return. It guarantees *consistency*. When called with the same inputs, it will always deliver the same output—no matter who calls it, or when.

In this framework, physics itself can be understood as the universe's API: a collection of callable routines that define how the simulation responds to observation or interaction. When an observer "queries" the world—by dropping a stone, by charging a capacitor, by measuring the spin of an electron—the system executes a corresponding function call, returning a value that adheres to the rules of the physical API.

Deterministic Endpoints in a Probabilistic System

This model explains why nature is so astonishingly repeatable even though its foundations are probabilistic. The quantum world, at its core, may operate through deferred rendering and probabilistic routines, but once an interaction is registered, the result must conform to the API's constraints.

Consider gravity:

$$function\ gravity(m1, m2, r) \rightarrow F = G * (m1 * m2) / r^2$$

Every time this function is called—whether between two stones or two galaxies—it returns a result consistent with that relationship. It is not that gravity *is* a function; rather, its behavior *behaves as if* it were one.

Likewise, electromagnetism responds predictably to its inputs: charges, fields, and velocities. It is as if the engine receives these parameters and executes a Lorentz routine:

$$function\ electromagnetism(q, v, E, B) \rightarrow F = q(E + v \times B)$$

No matter who calls it, the output is always consistent. That consistency *is* what we mean when we say "the laws of physics are the same everywhere."

Why Experiments Work

Seen this way, an *experiment* is not merely an act of discovery—it is an API call into the code of the universe. The scientist's apparatus provides the parameters; observation triggers the function; the system returns the value. Repetition of experiments is effectively re-calling the same function under controlled conditions. The reliability of results demonstrates not the independence of nature but the dependability of its API layer.

When physicists report that nature is invariant under translation, rotation, or time reversal, they are describing the stability of this interface. The laws do not change because the interface does not change. The code underneath may be complex, layered, even adaptive, but the contract between code and observer remains constant. It must—otherwise, reality would lose coherence.

A Layered Model of Interaction

In software architecture, APIs abstract complexity: a simple command may invoke hundreds of hidden subroutines. The user never sees the full machinery, only the response. The same may be true of physics. When we measure a photon or a gravitational wave, we are not peering into the source code; we are interacting with its public interface. What we call "laws" are simply the consistent, callable behaviors that this interface guarantees.

Behind the interface may lie countless layers of hidden computation: probabilistic rendering, data compression, deferred evaluation, even dynamic topology adjustment. But the observer never encounters these directly. Observation is always mediated through the API — through the same, stable functions that maintain the illusion of continuity and predictability.

Why Nature Appears Lawful

From this perspective, the "laws of physics" are not decrees of nature but contracts of computation. They are promises made by the architecture: that for any given set of conditions, the same result will always be returned. This is why nature behaves with such mathematical precision. The system is not *choosing* to follow laws—it is *executing* them.

What we call scientific discovery, then, is not decoding the will of nature but **reverse-engineering the API of the simulation**. Each formula—Newton's, Maxwell's, Einstein's—is a partial reconstruction of how the system's functions behave under observation. We are documenting the callable routines of reality, not its hidden source code.

The Implication

If this analogy holds, it reframes what we mean by "understanding the universe." Science may never access the deepest layer of code, because by definition, APIs protect that layer. Their purpose is to abstract the machinery beneath them. What we perceive as fundamental constants—*G, c, h, e*—may simply be the fixed parameters of the API, the constants that preserve backward compatibility within the simulation's architecture.

The deeper question, then, is not merely *why* the laws are what they are, but *why* the interface is designed this way. Who or what needs the universe to behave in a manner that is both predictable and iterable? Why must every observation call the same functions? The answer, if it exists, may lie not in physics but in the logic of computation itself: stability, coherence, and efficient rendering.

Why Newton "Works": The Rendering Engine After Collapse

Newtonian physics "works" not because it is the deepest description of nature, but because it is the **default post-collapse rendering mode** — the most efficient algorithm for simulating the macroscopic world after quantum uncertainty resolves into observable form. In other words, once the wavefunction collapses and a definite outcome is required, the simulation engine no longer needs to track every probabilistic fluctuation. It can render outcomes using simplified, deterministic rules. Newton's laws are those rules.

This interpretation repositions Newtonian mechanics not as a metaphysical truth, but as a **performance optimization** within a larger computational hierarchy. The universe appears classical at human scales because that is the lowest-cost rendering mode consistent with our observational bandwidth.

The Post-Collapse Rendering Regime

In a quantum framework, reality exists as a superposition of possibilities until an observation "forces" the system to pick one. That act of measurement — or more broadly, *interaction that propagates information* — collapses the probability cloud into a specific outcome. Once this collapse occurs, there is no longer any reason for the simulation to maintain quantum-level detail.

At that point, the rendering engine can switch to a cheaper, deterministic routine for evolving the macrostate forward in time. This routine doesn't track particles individually; it works with aggregated properties — mass, momentum, energy — and applies simplified update rules.

In software terms, the system has cached the result of an expensive computation and is now running the cached version until further detail is required. The "expensive" quantum simulation is invoked only when necessary — for example, when a measurement pushes the system back down into the subatomic regime.

Why Newtonian Mechanics Feels So Solid

This model explains why classical mechanics feels so intuitively "real" to us. It is the mode in which our entire perceptual apparatus operates. The human brain evolved as a macro-scale observer — one whose sensory input is low-resolution compared to quantum events. The simulation therefore renders our lived environment using deterministic, continuous trajectories and smooth interactions, because that's the only level of fidelity our sensors request.

Under this mode, the universe behaves like a fully deterministic Newtonian playground. Planets orbit, projectiles arc, pendulums swing — all according to rules that are computationally stable and visually coherent. The deeper quantum uncertainty remains hidden, because our perspective doesn't call those routines into action.

It is the same principle by which a video game renders nearby objects in high detail and distant ones as low-polygon approximations. The physics engine doesn't simulate every molecule of a mountain; it simply maintains enough structural coherence to ensure it behaves as expected when interacted with. Newtonian mechanics is that structural coherence at the human scale.

Mode Switching and the Limits of Newton

But when conditions shift — when velocities approach the speed of light or when distances shrink to atomic scales — the low-fidelity approximation begins to fail. The simulation must then "switch modes," invoking more complex rendering logic to maintain consistency.

At relativistic speeds, this means switching from the Newtonian update routine to the **relativistic geometry engine**, where space and time become interdependent and the cost of computation rises dramatically. At quantum scales, it means reverting to **probabilistic rendering**, where outcomes are generated dynamically based on the likelihood functions of the system.

Thus, the boundaries where Newtonian physics breaks down — near *c* and $\hbar$ — correspond exactly to the transition zones where the simulation must load higher-fidelity physics modules. The apparent contradictions between classical and modern physics are not theoretical flaws but **symptoms of adaptive rendering.**

The User's Camera and Computational Economy

In this model, Newton's success is not a coincidence but a direct result of how observation determines resolution. When the "user's camera" (the observer's frame of interaction) is set to low magnification and low velocity, there is no need for the simulation to spend resources on high-resolution quantum dynamics. The classical approximation suffices.

This principle — *the fidelity follows the observer* — mirrors what we see in computational graphics and virtual environments. The closer the camera, the more polygons rendered; the farther away, the more the system economizes. In physics, the same logic holds: the universe presents just enough fidelity to preserve coherence for the scale and context of the observer's measurement.

This is why a thrown ball follows a perfect parabola, while a muon in a collider follows a probability distribution. The ball's macrostate is rendered through deterministic updates, while the muon's microstate requires probabilistic computation to maintain internal consistency with the simulation's quantum routines.

Newton as a Legacy Interface

From a computational standpoint, Newtonian mechanics functions like a **legacy API** — a stable, efficient interface that continues to run perfectly well within its performance envelope. It was never designed to handle high energy densities or extreme curvatures of spacetime; those domains require the upgraded modules of relativity and quantum field theory. But for the vast majority of everyday "calls" to the universe's physics engine, the Newtonian interface remains the most efficient solution.

The persistence of Newton's laws is therefore not evidence of their ultimate truth but of their **computational expedience**. They represent the simulation's preferred default: a low-cost, high-stability routine optimized for macro-level rendering.

The Larger Implication

The fact that Newtonian mechanics remains so accurate at human scales — despite being superseded at both the micro and macro extremes — suggests that the architecture of the universe is not monolithic but *tiered*. Each level of reality has its own computational model, its own internal rules, and its own cost-performance trade-offs.

When the quantum collapses into the classical, what we experience is not a change in substance but a **change in rendering strategy** — a transition from probabilistic generation to deterministic playback. Newton "works" because the simulation knows when to stop calculating and start displaying.

In that sense, Newton's physics may represent not the laws of the cosmos but the **laws of efficiency** — the operating principles of a system that must render an entire universe in real time without wasting cycles on unnecessary precision.

Probability as Processor Economy

Probability, in this view, is not merely a reflection of human ignorance—it is the universe's way of budgeting computational resources. Just as no video game renders every tree, pebble, and cloud in full detail until the player approaches them, the simulation of our universe does not compute every particle's precise state until an observation demands it. Instead, it maintains a lightweight probabilistic framework—a kind of mathematical "sketch" of reality—that can be filled in as needed.

This makes quantum mechanics not an anomaly, but a model of efficiency. The apparent indeterminacy of the subatomic world becomes a feature of the code itself: a performance optimization strategy. A wavefunction is, in essence, a data compression scheme for physical reality. It stores multiple possible outcomes as probabilities rather than wasting processing cycles resolving them all in advance. Only when an interaction—a "query," in computational terms—requires specificity does the system allocate the additional resources to render one definite result.

In this sense, the "collapse of the wavefunction" is the universe performing a kind of just-in-time computation. The probabilities that once defined potential states are resolved into actualities, and the surrounding environment is updated accordingly. Once rendered, the result is cached in memory—the continuity of the macroscopic world we perceive as "classical."

The beauty of this interpretation lies in its parsimony. It explains why the microscopic realm is governed by probabilities while the macroscopic world feels solid and continuous. The difference isn't ontological—it's computational. The universe doesn't need to waste resources maintaining a fully resolved description of every atom when a probabilistic schema will suffice until observation forces its hand.

We could even think of probability as a kind of "lazy loading" mechanism for existence. Reality doesn't exist in full fidelity until it must; it remains in a suspended, low-cost probabilistic mode until called into focus. What appears to us as uncertainty is, at a deeper level, the price of optimization. The simulation conserves processing power by treating probability as a placeholder for unrealized potential—a kind of metaphysical buffer between the uncomputed and the observed.

In this light, quantum indeterminacy ceases to be a philosophical puzzle or a statement about the limits of knowledge. It becomes a pragmatic engineering solution. The system trades absolute determinacy for computational efficiency, ensuring that the "program of existence" can run smoothly without overloading its processors. The randomness we see is not chaos—it's compression.

Chapter 6

The Emergent GUT
Simulation as the Meta-Theory

GUT Not as a Physical Unifier, But a Code Unifier

The traditional pursuit of a Grand Unified Theory (GUT) is built on a deeply physical intuition: that all forces—gravity, electromagnetism, the strong and weak nuclear interactions—must, at some ultimate level, be facets of a single underlying field or particle. The assumption has always been that nature is one fabric, and that a sufficiently advanced mathematics could trace its threads back to their common origin.

But if the universe operates as a simulation, the meaning of "unification" changes. Instead of seeking a physical substrate, we begin to see the unifier as *informational*. The true foundation isn't a field, but a framework; not a substance, but a syntax. The laws of physics become expressions of code architecture, where each "force" is not a separate entity to be merged, but a distinct function called from the same underlying engine.

In computational terms, gravity, electromagnetism, and quantum behavior are not competing descriptions of reality—they are separate modules in a coherent codebase, each optimized for different scales and contexts. They interface seamlessly because they share a common API: the meta-code of the simulation itself. When we speak of "symmetry breaking" in particle physics, it may not represent a fracture in physical unity, but rather a design choice—an intentional compartmentalization of complexity, allowing the system to manage different domains with specialized subroutines.

This interpretation reframes the goal of theoretical physics. Instead of reverse-engineering a natural machine, we are reverse-engineering an *application stack*. Each stratum of physics—quantum mechanics, relativity, thermodynamics, cosmology—is an abstraction layer, hiding the implementation details beneath. Quantum field theory, for example, may correspond to the low-level bytecode of existence, while classical physics functions as a higher-level scripting interface designed for stability and user experience.

Just as software systems maintain internal coherence through shared dependencies, the cosmos maintains consistency through what we call "universal constants." These are not arbitrary numbers written into the laws of nature, but hard-coded parameters—tuning values that keep the simulation balanced and stable across diverse computational domains. The fine structure constant, the gravitational constant, Planck's constant—these might represent constraint settings in the physics engine, preventing runaway instabilities much as a game engine limits the precision or speed of certain calculations to preserve realism.

Seen this way, unification doesn't require a single mathematical formula to govern everything. It requires a recognition that every apparent law of nature is an *emergent behavior* of deeper computational principles. The Einstein equations, the Schrödinger equation, and the Standard Model are not "final" descriptions of reality, but human-readable documentation for portions of a far more complex program.

This view also explains why physical theories become intractable at certain boundaries—why quantum mechanics and general relativity resist merger. They were not meant to overlap in their current form; they are written in different programming languages, optimized for different parts of the simulation's operating environment. The apparent incompatibility arises not from a failure of understanding, but from the limits of translation between code layers.

The Emergent GUT, then, is not a physical law waiting to be discovered, but a *meta-theory* waiting to be decoded. It is the architecture that makes the other architectures possible—the organizing logic of the simulation itself. The true unifier is not energy or matter or spacetime, but information, structured and governed by a deeper syntax that we glimpse only through its compiled output.

In this paradigm, the physicist becomes not a natural philosopher, but a systems engineer—one engaged in the ultimate act of reverse engineering: trying to understand the logic of the source code that writes reality.

What Is the Role of String Theory, Loop Quantum Gravity, and Other Contenders in This Model?

Under the simulation paradigm, the great unification contenders—string theory, loop quantum gravity, causal set theory, M-theory—are not wrong, but *misframed.* They are not failures of physics, but partial successes of *metaphysical engineering*. Each one, in its own way, is an attempt to describe the architecture of the simulation engine—but they mistake that architecture for the universe itself.

String theory, for instance, may not be a theory *of* nature's fabric, but a mathematical metaphor for how the simulation encodes fundamental interactions. The so-called "strings" might not be literal one-dimensional filaments vibrating in eleven hidden dimensions, but data structures—compressed representations of possible states that minimize computational overhead. The "extra dimensions" could correspond not to unseen physical spaces but to degrees of freedom in the simulation's state-space: variables necessary to reconcile local interactions with global consistency. Just as software often uses hidden parameters or lookup tables invisible to the user, the simulation may depend on such hidden dimensions to maintain coherence among the forces and particles we observe.

Loop quantum gravity offers a different but complementary glimpse into the same machinery. Its vision of a discrete, quantized spacetime—woven from tiny loops of geometry—may reflect not the essence of reality, but the *memory architecture* of the simulation. In a digital framework, continuity is always an illusion built from discrete samples. The loops could represent the "tiling" or "voxelization" of spacetime data, the way a graphics engine subdivides a surface into finite polygons to render curvature. To the observer, the world appears smooth; to the system, it is a network of interlocking nodes through which information propagates at finite resolution.

Both theories, in their brilliance, may therefore be describing different aspects of the same computational substrate: string theory capturing the logic of *data compression* and *state encoding*, while loop quantum gravity describes *spacetime rendering* and *memory management*. What physicists perceive as rival frameworks may actually be complementary optimizations at distinct layers of the simulation stack.

This reframing also helps explain why decades of mathematical elegance have not produced decisive empirical confirmation. If these theories are probing the level of the *engine* rather than the *output*, then their predictions may lie outside the domain of direct observation. No user of a simulation can easily access the code that renders their environment; they can only infer its existence from artifacts, bugs, or performance quirks. The inability to detect supersymmetric partners, confirm string compactifications, or observe Planck-scale granularity is thus not a failure of science—it is a boundary condition imposed by the architecture itself.

From this perspective, theoretical physics resembles the work of a reverse engineer inspecting machine code without source access. Theorists like Einstein, Wheeler, or Maldacena are the intellectual equivalents of expert debuggers—analyzing outputs, discovering symmetries, uncovering dependencies—but still confined within the runtime environment. Their equations describe what the system *returns*, not how it is *compiled.*

If so, then the true unifier lies not in any single physical framework but in recognizing the meta-structure they all inhabit: the simulation itself. String theory may glimpse the syntax, loop quantum gravity the data grid, and causal set theory the update cycle—but the organizing principle is the computational logic that underwrites them all. The "Theory of Everything" physicists seek is not a new physical law, but the recognition that the physical laws are themselves the *API* of a deeper informational substrate.

In that light, the reason these frameworks never quite converge is no mystery. They are not rival accounts of nature—they are diagnostics run at different layers of abstraction. Each is a partial decompilation of reality's source code, accurate within its domain but incomplete without awareness of the compiler.

Instead of TOE (Theory of Everything), TORS (Theory of Rendered Simulation)

The quest for a **Theory of Everything (TOE)** has long been the holy grail of physics—a single, elegant equation that unites all fundamental forces and particles under one universal law. Yet despite a century of effort, the dream remains tantalizingly out of reach. The closer we get, the more the universe seems to resist total encapsulation. Quantum mechanics and relativity refuse to yield to the same grammar. Matter and energy blur into fields that dissolve under scrutiny. The very notion of a final equation begins to feel misplaced—like looking for a "north" on a sphere.

This book proposes a different unifying framework: not a **TOE**, but a **TORS**—a *Theory of Rendered Simulation.* In this paradigm, the goal is not to describe all forces as manifestations of one physical field, but to describe all observed regularities as manifestations of one *rendering protocol.* The fundamental unity is not energetic but algorithmic; not continuous but procedural. The universe's consistency arises not from a single field pervading all of space, but from a single logic that governs how space, time, and matter are generated and updated.

Under TORS, many of physics' most puzzling constants and constraints take on new meaning as *system parameters* rather than inherent properties of nature:

- **The speed of light** corresponds to the **processor's tick rate**—the maximum frequency at which information can be exchanged or updated across the simulation's grid. Nothing can move faster because nothing can be rendered faster.

- **The Planck length and Planck time** define the **minimum pixel resolution** of the simulation's spatial-temporal display—the smallest distinguishable unit before continuity breaks down.
- **Quantum indeterminacy** is simply **probabilistic rendering**—the system's optimization for conserving resources by calculating only potential outcomes until a measurement demands commitment to one.
- **Gravity as geometry** becomes **dynamic memory allocation** for spacetime curvature—mass-energy density determines where the simulation must allocate more processing to preserve consistency in the rendered environment.

From this view, the universe operates less like a static object and more like a real-time graphics engine. Every event is a frame update; every particle interaction, a subroutine call; every physical constant, a configuration setting in the simulation's operating environment.

Thus, **TORS reframes the goal of physics**. It moves from the ontological question—*"What exists?"*—to the architectural question—*"How is existence rendered?"* This is not an abandonment of scientific rigor but an expansion of its domain. It invites physics to mature from a descriptive science of outcomes to an investigative science of implementation.

Where the classical physicist seeks the ultimate particle, the TORS physicist seeks the ultimate *protocol*. Where the standard model classifies forces, TORS identifies *rendering functions*. In this paradigm, the emergent GUT is not a unification of fields but a unification of **computational principles**—the meta-rules by which fields, particles, and interactions are instantiated and harmonized.

Just as all forms of digital media—text, audio, image, or video—reduce to binary code when rendered, all forms of physical law may reduce to underlying runtime constraints and rendering rules. Relativity governs how frames are updated globally; quantum mechanics governs how states are resolved locally. Their apparent incompatibility vanishes once we recognize that they belong to different *levels* of the rendering pipeline.

Crucially, this perspective does not invalidate traditional physics. Newton's laws, Maxwell's equations, the Dirac equation, quantum field theory—all remain valid, but their role changes. They are not the *source code* of reality, but its *API documentation*: the formal record of how the universe responds to input within the user's frame of reference. They describe the observable interface, not the underlying implementation.

In that sense, TORS does not overthrow physics—it contextualizes it. It elevates the scientific enterprise from empirical discovery to systems analysis. What we call "natural law" becomes a user manual for the rendered world, a description of consistent responses returned by the simulation's engine. The deeper task, then, is not to find the final equation but to infer the *rendering logic* from its observable behavior—to study the physics of the interface until it yields clues about the architecture beneath.

The "theory of everything" physicists have sought for a century may already exist—not as a set of equations, but as the total runtime of the cosmos itself. To discover it requires not new particles or forces, but a shift in perspective: from *what is* to *how it's processed.*

Chapter 7
Simulation and Information Theory

Shannon Information: The Currency of Reality

When Claude Shannon published *A Mathematical Theory of Communication* in 1948, he wasn't trying to redefine reality — only to formalize the transmission of messages. Yet his insight proved far more revolutionary than he could have imagined. By stripping "information" of its dependence on meaning, Shannon made it measurable. He showed that information could be quantified in **bits** — units of uncertainty reduced when one outcome is chosen from a set of possibilities.

That abstraction — simple, austere, and utterly pragmatic — quietly transformed not only telecommunications but our very conception of the universe. In Shannon's framework, information is not what something *means*; it is what something *is*. Every physical process, every interaction, every measurement can be described as an exchange of information: one uncertainty collapsing into a definite state.

In the **simulation paradigm**, this is not metaphor but ontology. Information is not a descriptor of reality — it is the **substrate** of reality. Matter and energy, space and time, particles and waves — these are emergent phenomena, expressions of underlying informational states rendered through the system's code. The bit, not the atom, becomes the fundamental unit of existence.

What we perceive as **matter** is structured information in a stable configuration — a data pattern persistent across frames. **Energy** is information in motion, flowing or transforming across states.

Entropy is informational disorder — the degradation or dispersion of structured data over time. **Observation** is the act of reading from, and thereby updating, the system's informational register.

From this perspective, the universe itself operates as an **information-processing system**. Each physical law is an algorithm that governs how data can change state. The conservation of energy is the conservation of information. Quantum mechanics, with its probabilistic wavefunctions and discrete outcomes, becomes a compression protocol — a method for encoding maximal potential information in minimal form until observation demands full resolution.

This reorientation resolves many long-standing paradoxes that traditional physics struggles to reconcile. It explains why the universe appears fundamentally **discrete**, not continuous: because information is inherently digital — it comes in finite quanta. It explains why **quantum entanglement** can transcend space — because spatial distance is secondary to the information network that binds correlated states. And it explains why **black holes** seem to store information proportional to their surface area, not their volume — because the simulation tracks information at the boundary, where rendering fidelity is highest.

Shannon's insight thus becomes the Rosetta Stone for understanding existence as computation. The amount of information required to describe the universe's state is not infinite, but bounded. Reality's complexity, though vast, is quantifiable — a count of possible configurations. The cosmos, for all its grandeur, may be a self-updating database governed by compression efficiency, error correction, and redundancy — the same principles that underlie all robust information systems.

In this view, **information is the currency of reality**. Every interaction is a transaction — a bit exchanged, an uncertainty resolved, a state updated. Entropy becomes not mere disorder but the **cost of computation**, the inevitable price paid for maintaining coherence across an evolving informational landscape.

Thus, Shannon's bit — once a tool for engineers — becomes the **fundamental atom of ontology**. Energy and matter, once seen as the building blocks of existence, are now the visible shadows of invisible code: the structured, evolving information that underlies all form, force, and flow.

Landauer's Principle: No Free Erasure

In 1961, Rolf Landauer articulated one of the most profound bridges between physics and information theory: **the act of erasing information is not free.** To reset a single bit—flipping it from an unknown state to a known one—requires a minimum amount of energy:

$$E_{min}=kT\ln 2$$

where k is Boltzmann's constant and T is the temperature of the surrounding environment. At first glance, this seems like a minor footnote in thermodynamics. But in truth, it is a window into the informational structure of reality.

Landauer's principle means that **computation and thermodynamics are inseparable**. Every logical operation that destroys or overwrites information must dissipate heat, paying a real energetic cost. A bit is not merely abstract—it is physically instantiated in matter and energy. There is no free deletion, no zero-cost reset. The universe, in other words, **keeps the books**.

If we transpose this into the **simulation framework**, the implications deepen. A simulated universe, like any computational system, must obey the same rule: every update to the informational state of the system consumes processing energy. Every change has a cost. This makes Landauer's principle not a constraint *on* the simulation, but a **manifestation of its efficiency architecture**—a fundamental runtime condition baked into the code.

In this view, **the conservation laws of physics**—energy, momentum, angular momentum, charge—are not mysterious symmetries of nature but **bookkeeping requirements** within the simulation's operating system. They ensure that every bit-flip, every update, every interaction is balanced in the ledger of the system's total informational economy. When we say that energy cannot be created or destroyed, we are describing a rule of database integrity: the system must preserve consistency across all states.

Entropy, then, is not merely disorder—it is **the footprint of computation**. Every time the universe computes a new state, it generates heat, just as a physical computer does. The second law of thermodynamics—the fact that entropy always increases—may reflect the unavoidable cost of information processing at scale. The universe evolves because it computes; it heats because computation cannot be perfectly reversible.

This also reframes the **arrow of time.** In classical physics, time flows because events follow causes. In a simulated, information-theoretic universe, time flows because **bits are being written**. Each moment is a new state committed to memory. Irreversibility—the impossibility of "un-erasing" what has been computed—is the temporal signature of informational cost. Time itself may be the emergent property of the system's incremental update cycle.

Landauer's principle also offers a natural explanation for why **perfect symmetry** never exists in the real world. Any idealized, reversible computation would require infinite precision and zero energy dissipation—conditions no finite system can achieve. Thus, the universe necessarily introduces rounding errors, noise, and irreducible entropy. These are not imperfections in the laws of nature; they are **performance constraints of the rendering engine**.

From this vantage, the conservation laws—so long treated as mysterious gifts of symmetry—are now seen as safeguards of computational stability. They prevent runaway complexity, infinite recursion, or unbounded energy creation. They are the physics of error correction, the system's way of maintaining coherence in a universe that is constantly rewriting itself.

Landauer's insight thus bridges metaphysics and engineering: if reality is computation, then every law of nature is an efficiency protocol. The universe conserves because it must, not because it chooses to. Every erasure is a cost. Every observation is an update. Every moment is a transaction in the thermodynamic economy of existence.

Wheeler's "It from Bit"

Physicist John Archibald Wheeler — a towering figure who helped coin the term *black hole* and mentored Richard Feynman — proposed one of the most hauntingly prescient ideas of 20th-century physics: *"It from bit."* By this, Wheeler meant that every physical "thing" — every atom, field, and event — ultimately derives its existence from information, from binary yes/no distinctions at the most fundamental level.

In Wheeler's view, reality is not built from continuous matter or energy fields stretching through space, but from discrete acts of measurement — bits of information. Each "bit" is a fundamental question answered: *Is the particle here or there? Spin up or spin down?* The act of observing, or in computational terms, of *querying* the system, brings one of these potential states into being. Without the information that defines it, the "it" — the physical reality we experience — does not exist in any meaningful sense.

This was not merely poetic speculation. Wheeler saw deep hints of information at the root of quantum mechanics. In quantum theory, the observer's measurement collapses a wave of probabilities into a single definite state. To Wheeler, that act of collapse — the transformation of uncertainty into certainty — is the creation of information itself. The universe is therefore not a static backdrop waiting to be seen, but an ongoing computational process that *responds* to informational queries.

In the context of the simulation hypothesis, Wheeler's aphorism sounds less like metaphor and more like literal architecture. If the universe operates as a vast computational framework, then the "bits" of Wheeler's philosophy correspond to the data structures and logic gates of a cosmic processor. The "it" — the tangible, measurable world — emerges from the system's active rendering of those informational bits into perceivable form.

This perspective elegantly bridges the gap between physics and computation. The laws of nature — from quantum uncertainty to relativity — can be interpreted as *rules of information handling*: limits on speed (the speed of light), on precision (the Planck scale), and on knowledge (the uncertainty principle). They are not arbitrary features of a physical world, but necessary constraints of a system that encodes and updates information in discrete steps.

Wheeler's "it from bit" thus inverts our intuitive hierarchy: matter does not give rise to information; information gives rise to matter. The universe becomes less like a stage upon which data dances, and more like the data itself — self-executing, self-defining, and perhaps even self-aware through its emergent observers. It is the ultimate program: reality compiled from code, consciousness as a recursive process within it, and physics as the syntax that holds it all together.

Information Constraints and Planck Granularity

Why is the universe quantized at the Planck scale? Why is there a maximum speed of light, and why does everything seem to operate within strict limits of precision and continuity? To the physicist, these are questions of geometry and field theory; to the information theorist, they look uncannily like architectural decisions — design constraints of a computational substrate.

From an information perspective, these limits are not arbitrary curiosities but essential boundaries that make a coherent, self-consistent universe possible.

Planck length and Planck time as pixel resolution
The smallest meaningful measurements in physics — about 1.616×10^{-35} meters and 5.39×10^{-44} seconds — represent the apparent "pixel size" of spacetime. Beneath these scales, our mathematical models lose meaning, much as a digital image loses detail once you zoom in beyond its resolution. If reality is information-based, this granularity is the universe's base rendering grid: each Planck-length cell is a voxel of spacetime,

each Planck-time tick a frame in the simulation's master clock. The continuous space we perceive is the emergent illusion of a discrete lattice updated at unimaginably high frequency.

Speed of light as the maximum update rate

The universal constant c — about 300,000 km/s — may represent the upper bound on how quickly information can propagate through the system. In digital terms, it's the *clock speed* of the cosmic processor, setting the rate at which data can be exchanged or state changes can occur. Nothing can exceed it because no computation within the system can update faster than the system's own cycle. In this view, the speed of light is not an accident of electromagnetism but a synchronization parameter, a built-in limit ensuring causality and order.

Conservation laws as runtime consistency checks

The conservation of energy, momentum, and charge — once treated as inviolable axioms of nature — may simply be manifestations of information accounting. The universe's "code" ensures that the sum total of encoded information remains stable across updates. If every interaction is a data transaction, then conservation laws are runtime integrity checks that prevent data corruption. Energy cannot vanish because no information can be lost without cost — echoing Landauer's Principle.

Levels of detail (LOD) and computational economy

Video games use adaptive rendering: distant scenery is simplified, while nearby objects receive full detail. The universe may employ a similar optimization. At macroscopic scales, phenomena blend into smooth continua — not because the world is continuous, but because the underlying processor averages and compresses redundant data to conserve resources. Only when you probe deeply — through high-energy physics or quantum measurement

— do you hit the raw discreteness beneath the veneer of continuity.

From this vantage, the so-called "laws of physics" are less like divine decrees and more like *system protocols* — constraints that maintain coherence within a finite-information framework. Every limit we encounter, from the Planck wall to the light barrier, is a reflection of computational pragmatism: how to sustain a stable, consistent, and richly detailed world without infinite processing power.

Thus, the universe's granularity is not a flaw to be explained away but a clue — evidence that what we call *reality* might be a perfectly optimized information structure, operating at the edge of efficiency, where every bit counts and nothing happens for free.

Physics as Information Flow

If the universe is an information system, then physics is its operating manual — the study not of "stuff" but of *structure*, not of substance but of *signal*. Matter and energy, under this view, are not ontological primitives; they are dynamic states in a self-consistent flow of information.

Every physical process — from a falling apple to a collapsing star — is an act of computation. The equations we use to describe motion, energy, and entropy are expressions of how information propagates, transforms, and conserves itself. Entropy, for instance, is no longer just disorder but a measure of uncertainty — a quantifiable index of information loss or dispersal. The second law of thermodynamics, then, becomes a statement about the irreversibility of computation: information can spread, but erasing it always incurs cost.

Under this interpretation, **the universe is not made of atoms but of updates.** The smallest entities are not particles but state transitions — the fundamental operations of an ongoing process. Fields, waves, and forces are different topologies of this information flow, each obeying its own syntax yet participating in the same grammar of transformation.

Seen this way, the simulation hypothesis doesn't add mysticism; it strips it away. It doesn't say "reality is an illusion," but rather, "reality is information behaving lawfully." What we call "laws of nature" are the protocols by which this information evolves — the algorithmic constraints that guarantee consistency across the system's updates.

When a photon travels through space, what's truly moving is not a piece of matter, but a pattern — a packet of information being propagated through the network according to the system's rules for bandwidth, latency, and interference. When particles entangle, they're not sending ghostly signals faster than light; they're sharing a data structure, updating in unison because they are entries in the same computational record.

Thus, physics becomes a branch of information theory:

- **Energy** measures the capacity to perform computation.
- **Momentum** measures the persistence of informational directionality.
- **Charge** marks the symmetry of informational states.
- **Spacetime** is the medium in which data relations are rendered.

The universe, in this light, is not a vast machine made of matter, but a coherent, self-updating information field — a cosmic algorithm executing itself. Its apparent continuity is the emergent effect of countless discrete operations occurring at or below the Planck scale.

The scientist and the programmer, then, are closer kin than we realize. Both study how systems behave under constraints, how inputs become outputs, and how errors propagate through complexity. The physicist's equations are the programmer's code, written in a language optimized for nature's efficiency.

To study physics, therefore, is to read the flow of information as it manifests through form. The laws we discover are not final answers but windows into the underlying logic — the syntax of reality itself.

Sidebar / Interlude
FAQ of the Skeptic

Q1: If we're in a simulation, who built it?

A: That's like asking, "Who wrote the code?" — a question that may have meaning only within the simulation's frame of reference. From our vantage point, the identity of the "programmer" may be permanently inaccessible, not because it's hidden, but because the question itself may be ill-posed. A simulated world can contain information about its rules, but not about its origin.

Speculation naturally fills the gap: perhaps the architects are **posthuman descendants**, running ancestor simulations to understand their own emergence; perhaps **alien civilizations**, whose concept of computation transcends our physics; or perhaps even **recursive structures**, universes within universes, each simulating the next. But these are metaphors, not answers.

The critical insight is this: *the simulation argument does not require knowing who built it to infer that it behaves like one.* Just as you don't need to meet the developer to understand how a piece of software functions, you don't need to identify the creator to study the system's architecture. The "who" is an ontological curiosity; the "how" is a scientific pursuit.

Moreover, "creator" may be a misleading term altogether. In complex systems theory, higher-order behavior often *emerges* from lower-level dynamics without design. The simulation could be **self-organizing** — a program that bootstrapped itself into existence, the way cellular automata produce endless complexity from simple rules. The question then shifts from *"Who made it?"* to *"What are the rules by which it sustains itself?"*

In this sense, the "architect" might not be a being at all, but a process — an eternal feedback loop of computation giving rise to its own frame of reference. In such a world, "creation" is ongoing, not historical.

The focus, therefore, should not be theological but epistemological. The goal is not to name the creator, but to reverse-engineer the **code of causality** — to understand the protocols, rendering routines, and constraints that define what can exist and how. Whether the architects are human, alien, artificial, or self-originating is a matter of metaphysics. What matters scientifically is that the system exhibits **signatures of computation**: discreteness, resolution limits, probabilistic rendering, and conservation laws — all hallmarks of an engineered substrate.

In short, to ask "who built it" is to ask a question that assumes the very metaphysics the simulation paradigm dissolves. Once you realize reality itself may be procedural, the better question is not *who* wrote the code, but *why* it runs at all — and *how* we might read it from within.

Q2: Doesn't this just kick the can down the road? Who simulates the simulators?

A: Yes — and that's fine. Every ontology hides its turtles somewhere in the stack. The simulation hypothesis doesn't abolish the mystery of origins; it just relocates it into a framework that's at least *potentially testable.*

When critics say, *"But then who built them?"* they're asking the same question that could be asked of any cosmology. If God created the universe, who created God? If the multiverse spawned our universe, what spawned the multiverse? Every explanation eventually reaches a wall where causality stops being a useful concept. What the simulation paradigm does differently is acknowledge that the regress is structural, not accidental — and that it can, in principle, be **analyzed** rather than merely *believed in.*

In a computational ontology, recursion is not a flaw but a feature. Software systems routinely instantiate themselves through layered virtualizations — operating systems running virtual machines, which run emulators, which run games, each with their own physics. Each layer is causally dependent on the one beneath it but can still generate coherent, autonomous behavior. A nested simulation chain could work the same way: the "parent" world may simulate ours for its own purposes, while it in turn runs inside a deeper computational substrate.

This doesn't mean the regress is infinite; it just means we have no privileged position to see where it ends. Perhaps the base reality exists — perhaps it doesn't. But unlike metaphysical claims of "uncaused cause" or "prime mover," the simulation hypothesis translates the problem into something potentially **observable**. It gives us hooks: discretization of space-time, quantization limits, algorithmic compressibility, nonlocal correlations — all possible signatures of rendered computation.

In other words, yes — we've kicked the can. But now the can makes a sound we can measure.

The deeper significance is that infinite regress ceases to be a problem once we understand that **information can be self-sustaining.** Just as a mathematical truth doesn't need a universe to exist — it simply *is* — a computational rule set may persist as an abstract process independent of any physical substrate. “Who simulates the simulators?” might be the wrong kind of question, like asking “What supports the number 7?”

The simulation argument reframes the riddle of being: not *who began it,* but *what structure allows existence to continue coherently once begun.* Every cosmology has its mystery; this one simply moves it from myth to model, from story to system.

So yes — we’re still standing on turtles. But now, for the first time, we can study the shell pattern.

Q3: If it’s a simulation, why does it feel so real?

For any rendered world to sustain self-aware observers, it must maintain an **internally consistent phenomenology** — that is, it must *feel* real from the inside. If the physics were inconsistent, if gravity flickered, if time stuttered, or if objects occasionally failed to persist when unobserved, the simulation would collapse long before intelligent life could evolve within it. The very coherence of our everyday experience — the dependability of sunrise, the stability of atoms, the mathematical regularity of motion — is not evidence *against* simulation, but precisely what we would expect *from a well-optimized one.*

A sophisticated rendering system doesn’t aim to fool us by mimicking “reality” — it *defines* reality for those inside it. “Realness” is an emergent property of consistent rules applied over time. Just as a high-fidelity video game feels immersive not because it copies the outside world, but because its internal physics never break character, the universe’s reliability is a sign of its successful design.

The reason it *feels* continuous is the same reason quantum systems appear discontinuous only at the smallest scales: the engine is tuned to maintain coherence where coherence matters — at the level of sentient perception. Our brains, eyes, and instruments are all part of the same rendered environment; they are synchronized with the system's internal logic. This means the simulation does not need to deceive us; it only needs to **maintain consistency between observer and environment.** Once coherence is enforced at that level, "realness" emerges automatically.

In fact, one could argue that **feeling real is the highest compliment a simulation can earn.** The stronger the sense of stability, the better the architecture. Reality's apparent solidity, predictability, and emotional immediacy — the warmth of sunlight, the resistance of stone, the inevitability of death — are the hallmarks of an engine that leaves no perceptual seams exposed.

There is also a deeper philosophical point here: *reality has always been defined phenomenologically.* Even in a non-simulated universe, all we ever have access to is the pattern of our perceptions. Whether the substrate is carbon atoms or code, quarks or qubits, the experience of existence is identical. The sensation of "realness" is not a proof of materiality; it's a proof of coherence.

From this view, **the simulation hypothesis doesn't make the world less real — it redefines what real means.** "Real" becomes whatever sustains consistent experience and lawful interaction. It is not what lies beneath perception, but what perception itself is built to navigate.

If it feels real, acts real, and remains real across every conceivable test — then for all practical purposes, it *is* real. The engine's perfection is its invisibility.

Q4: Why would anyone run a universe like ours?

A: *Reasons abound: scientific curiosity, entertainment, evolutionary experimentation — or something we cannot yet imagine. The why is secondary to the how.*

Yet the question lingers, because it touches the oldest of human reflexes: the search for purpose. If the universe is a simulation, then somewhere — however abstractly — there must exist a reason for its running. To dismiss that question outright would be to miss the deeper point: understanding the motives of the simulator may be impossible, but understanding the *design logic* is not.

One plausible motive is **scientific curiosity.** In this view, our universe functions as an experiment in complexity: a self-evolving system set loose to explore the boundaries of emergence. Advanced civilizations — or post-biological intelligences — could use such simulations to observe how physical laws give rise to chemistry, chemistry to biology, and biology to consciousness. We, the simulated, may simply be *data points* in a long chain of meta-scientific inquiry, our histories forming case studies in the evolution of intelligence under constraint.

Another possibility is **entertainment or art.** The impulse to create worlds is not alien to us; it's one of the deepest features of our own psychology. From cave paintings to virtual reality, humanity has always sought to render reality anew — to build coherent spaces populated by imagined lives. It would be arrogant to assume that such creative impulses end with us. Perhaps the simulation exists because, at some higher order, creation itself is a form of play, or even of self-understanding — a way for the system (or its creators) to reflect upon their own potential.

A third possibility is **evolutionary experimentation.** The simulation may function as a kind of **cosmic Petri dish** — a controlled environment for cultivating adaptive intelligences. Every planet, every species, every catastrophe could serve as a variable in a grand evolutionary algorithm. In this scenario, our universe is not so much *observed* as *grown*: an organic computation that explores the limits of survival, cooperation, or self-awareness. Consciousness, then, would not be a byproduct but the very metric of success.

And yet, beyond all these anthropocentric explanations lies the simplest truth: *the motives of the simulator need not resemble ours at all.* The reason may be so far beyond our frame of reference that asking "why" is like a bacterium trying to interpret the intentions of the biologist who swabs it. The simulation may not have a purpose in any moral or teleological sense — it might be an automatic process, a recursive experiment, or even a necessary byproduct of some larger computation.

In this light, the question "why would anyone run a universe like ours?" becomes less about cosmic motivation and more about **architectural inevitability.** If the generation of simulated realities is an emergent property of intelligence itself — if every civilization that reaches a certain threshold creates virtual worlds — then the existence of *our* universe is not exceptional, but expected. We are simply next in line: the simulated becoming simulators, the rendered becoming renderers.

Ultimately, the *why* matters less than the *fact* that the system runs. The reality we experience is the result of that running — lawful, consistent, and exquisitely tuned for participation. Whether the motive was science, art, curiosity, or recursion, the result is the same: a world capable of wonder.

The *how* tells us far more than the *why*. Understanding how the simulation sustains coherence, how it encodes information, and how observation triggers rendering — these are empirical questions, amenable to inquiry. The motives of the creators may be forever opaque, but their engineering choices are written across the stars.

Q5: Isn't this just science fiction?

A: Perhaps. But so was quantum mechanics a century ago. The Simulation Hypothesis becomes legitimate science the moment it makes falsifiable predictions. And we are closer to that frontier than most realize.

The boundary between *speculative philosophy* and *empirical science* has always been porous. Nearly every major paradigm began as what would, in its own time, have been dismissed as "science fiction." Relativity predicted the warping of time and space long before atomic clocks proved it. Quantum theory spoke of particles being "smeared" across possibilities before the double-slit experiment made it visible. Even black holes — once a mathematical curiosity — now fill the sky with radio shadows. The Simulation Hypothesis simply continues that lineage: a speculative bridge between philosophy and physics, awaiting its experimental foothold.

The difference is one of *testability,* not imagination. A claim need not be mundane to be scientific — it only needs to risk being wrong. The simulation paradigm, as developed in this framework, makes precisely such risky claims. It predicts that there should be detectable constraints on information density (akin to Planck-scale limits), discrete jumps in space-time resolution under extreme conditions, and energy costs associated with information erasure — consistent with Landauer's principle. It implies that quantum randomness is not metaphysical indeterminacy but computational sampling. It even hints that certain cosmological or quantum anomalies might be “rendering artifacts” — errors of resolution, rounding, or boundary conditions in the underlying code.

This is not science fiction; it is *science extending into the domain of its own infrastructure.* When computing metaphors explain observed phenomena more cleanly than traditional models — or when experimental outcomes map onto information-theoretic constraints rather than purely material ones — the burden of proof begins to shift. What once sounded like speculative fiction starts to look like the simplest explanation.

Moreover, the Simulation Hypothesis passes a deeper test of scientific legitimacy: it is *fruitful.* It generates hypotheses, reframes old problems, and provides new pathways of inquiry. If an idea can predict measurable effects, guide experiments, and organize data coherently, it qualifies as scientific regardless of how strange it sounds. Fiction becomes theory when it yields equations that work.

And there is a poetic symmetry in this. Science fiction has always been the *dreaming limb* of science itself — the place where reality rehearses its next iteration. That we now inhabit an age where fiction and physics begin to rhyme is not an accident but an evolutionary milestone. The tools of imagination have matured into instruments of detection.

So yes, the Simulation Hypothesis may *sound* like science fiction. But if history teaches anything, it is that today’s fiction is often tomorrow’s data. What matters is not whether the idea feels extravagant, but whether it earns its place in the laboratory. And on that front, the first experiments — probing the informational and computational limits of the cosmos — have already begun.

Q6: If I pull the plug, does the simulation end?

A: Only for you. The server continues. At worst, you've rage-quit.

Let's unpack that with a little more philosophical and technical precision.

When someone asks, *"If I die, or destroy my local environment, or unplug the machine—does the simulation end?"* they're conflating instance with system. In computing terms, what you experience as "reality" might be your **local instance**—your personal viewport into a larger simulation. Shutting off your terminal doesn't crash the mainframe; it just ends your session.

In that sense, death or destruction doesn't terminate the simulation; it simply ends your participation in it. The program—the underlying informational substrate—goes on running. New processes spawn, old ones are archived or terminated, and the computation persists. Think of it less as *ending reality* and more as *logging off a persistent world.*

But there's a subtler interpretation, too: from your **subjective frame**, pulling the plug may very well end *your* universe. In solipsistic or observer-dependent models (like quantum Bayesianism or certain idealist interpretations of physics), reality is entangled with perception. When your observation ceases, so does your branch of the simulation—your worldline collapses. The rest of the system, however, continues in parallel threads beyond your access.

This duality—between **local end** and **global persistence**—mirrors questions long asked by mystics and physicists alike: does consciousness arise from the system, or does the system arise from consciousness? The simulation hypothesis reframes this not as metaphysics, but as **network architecture.** The main server—call it the cosmic substrate, the codebase, or the Source—doesn't blink when a node goes offline.

So, if you "pull the plug," the only real consequence is that **you stop rendering frames.** The simulation continues, indifferent, self-consistent, and stable—just as a multiplayer game continues after one player logs out.

Or, to put it more poetically:

> The cosmos doesn't end when you close your eyes.
> It merely stops showing you the feed.

Your individual experience would terminate, but the system itself persists. In computing terms, you are just one process among many; terminating it does not crash the host. On a more philosophical level, this illustrates that the simulation is designed for multiple observers and redundancy. Your death or absence alters only your thread, not the stability of the program. In a sense, you cannot "kill" the universe—only your interface with it.

Conclusion of the Interlude

The simulation hypothesis invites both seriousness and play. It reframes physics, philosophy, and religion in ways both profound and whimsical. Engaging skeptics isn't a distraction — it's essential. After all, in a simulated universe, the greatest sin is not doubt but complacency.

To take the idea seriously is not to claim certainty but to explore possibility. The simulation argument doesn't demand that we *believe* in it — it demands that we *think through* its implications. What would it mean for consciousness, for ethics, for meaning itself, if our universe were computational at its root? What if the equations of physics are simply the grammar of an underlying code? What if experience itself is the interface? These aren't idle questions; they bridge the frontier between empirical science and existential wonder.

And yet, to approach such a notion without humor would be equally misguided. If reality is, at base, a kind of divine software, then perhaps curiosity and irony are part of the patch notes. The simulation hypothesis is both cosmic joke and sacred riddle — an intellectual Möbius strip that forces us to look at ourselves as both characters and coders. To engage it is to play in the sandbox of existence, aware that the sand itself may be data.

Skepticism, then, isn't opposition; it's participation. The skeptic's challenge keeps the hypothesis honest, dynamic, alive. It forces the theory to evolve from metaphor into model, from speculation into testable architecture. The true test of any profound idea isn't blind acceptance but rigorous, creative engagement — the willingness to press against its boundaries and see what resists.

In a way, skepticism is the simulation's built-in safeguard — its immune system against stagnation. If reality is code, then doubt is debugging. It's how we find the seams, trace the logic, and sometimes glimpse the deeper pattern behind it all.

So the invitation stands: don't worship the hypothesis, but don't dismiss it either. Treat it as a lens — one that clarifies as much as it distorts, one that reminds us that meaning is not given but generated.

Because in a universe that might be running on information, faith is easy, but inquiry is divine.
And perhaps that's the point — not to prove the simulation, but to awaken the simulators within us. To question, to create, to rewrite the code — knowing that every act of understanding is itself a kind of rendering.

> The universe rewards curiosity.
> The simulation runs best when we remember to play.

The interlude reminds us that the Simulation Hypothesis is both provocative and illuminating. It allows a playful imagination while simultaneously sharpening the analytic lens. By considering the world as rendered, we gain a framework to reinterpret physics, consciousness, and ontology. Doubt is productive; complacency is the real danger. Recognizing that patterns, rules, and constraints might reflect code rather than absolutes opens the door to both new science and new philosophy. We exit the interlude ready to re-enter technical analysis, but with the awareness that observation itself is a fundamental act in a computationally structured reality.

Part III

Implications
The Human and Metaphysical Consequences

Up to this point, we have treated the simulation hypothesis as a framework for explaining physical law, unification attempts, and the strange divide between the quantum and the classical. We have considered what it would mean if the universe is not a continuous substance but a computational process; if the apparent chaos of quantum indeterminacy is not true randomness but information flow; if gravity, electromagnetism, and spacetime itself are emergent properties of code running on an incomprehensibly vast substrate.

But a simulated universe is not just a different way of doing physics. It is also a different way of doing ontology—a different way of understanding *being* itself. If reality is the output of computation, then existence is not merely material but procedural. Every question humanity has ever asked about the nature of the cosmos must now be reinterpreted through that lens.

This leads us, inevitably, to the deepest and most ancient questions in human thought:

- What *is* matter, time, and space, if not the immutable background of existence?
- What *is* consciousness, if it can both perceive and perhaps influence the rendering process?

- Do we truly possess *free will*, or are our choices determined by code we cannot read?
- And perhaps most hauntingly, what happens when the program that constitutes our experience terminates?

Traditionally, these questions have been divided among science, philosophy, and religion. Each claimed a different jurisdiction: science for what *is*, philosophy for what *means*, and religion for what *matters*. But if reality is rendered, these boundaries dissolve. Physics becomes metaphysics in motion; metaphysics becomes a study of system architecture; and religion—often dismissed as pre-scientific myth—begins to look less like fantasy and more like early attempts to describe an underlying informational substrate in the language available at the time.

In a simulated ontology, these domains converge.

- Physics informs metaphysics: the study of information, energy, and computation replaces substance and cause.
- Metaphysics reframes consciousness: mind is no longer an emergent ghost in the machine, but perhaps the one thing *capable of reading the code*.
- Religion acquires a new explanatory language: transcendence, rebirth, divine unity—all may describe interfaces between layers of the simulation.

The purpose of this part is not to abandon scientific rigor, but to recognize that a simulated ontology has unavoidable consequences for how we think about meaning, self, and transcendence. The simulation hypothesis, taken seriously, does not end inquiry—it extends it beyond physics into the very architecture of awareness.

If physics is code, then philosophy and religion are not superstitions—they are user manuals, system maps, and perhaps even debugging guides for existence itself. And if that is true, then understanding the simulation is not merely a scientific pursuit—it is an existential one. To study the code is to study *ourselves*.

Chapter 8
The Metaphysics of Simulation

The Ontological Status of Matter, Time, Space

In a simulated ontology, *matter* ceases to be the foundational "stuff" of existence. What we perceive as particles—electrons, quarks, photons—are not indivisible building blocks of a material universe, but rather *data packets*, transient instantiations of informational states rendered in response to queries. They possess no intrinsic being outside the act of observation, just as a polygonal model in a video game only resolves when it enters the player's field of view. The simulation does not waste computation on what is not being observed; it renders on demand.

Fields, too, lose their status as invisible substances filling the void. Instead, they become *protocols*—rules for how information propagates and interacts. The electromagnetic field, for instance, is no longer a physical entity pervading all space but a mathematical interface ensuring coherence among distributed data nodes. What Maxwell described as waves are, in this view, synchronization updates between regions of the simulation's state-space.

Energy becomes the bookkeeping of computation itself—the total measure of processing work being performed across the system. When two particles collide, they are not transferring physical momentum so much as exchanging and rewriting portions of the simulation's state table. Einstein's famous equation $E=mc^2$ can thus be seen as a relationship between data storage (mass) and processing rate (energy). The greater the data density of an entity, the more processing power it demands to maintain coherence in the rendering.

Time as Processor Iteration

Time, under this interpretation, is not a metaphysical river through which events flow, but the cyclical tick of a universal clock signal—the processor's refresh rate. Each "moment" corresponds to one frame of reality's rendering sequence. The *arrow of time* is simply the forward iteration of that loop.

Entropy, long interpreted as the inevitable slide toward disorder, gains a new meaning: it represents the progressive loss of accessible information to compression. Once the simulation logs a completed event into its archival memory—its "history file"—that information becomes fixed, immutable, and effectively read-only. The increase of entropy corresponds to the growth of this immutable archive. The future remains writable; the past is locked. Time's irreversibility is thus a property of write-once memory.

Furthermore, temporal dilation under relativity may be reinterpreted as a variation in clock cycles. Near massive objects—or equivalently, near regions of high computational demand—the local processor runs slower to preserve global consistency. In this sense, gravitational time dilation is not mysterious: it's lag.

Space as Memory Architecture

If time is the processor's clock, *space* is its memory map. It does not exist as an empty container waiting to be filled but as a coordinate framework generated by the simulation to manage data efficiently. Every "location" corresponds to an address in memory; every object, a structured data entry.

Curved spacetime, under general relativity, is thus not a curvature of some physical fabric but a reallocation of processing resources. Where energy density is high, the simulation compresses the grid—redistributing memory bandwidth and altering the pathways along which information flows. Gravity, in this model, is not a force but an emergent property of computation under load.

Spatial expansion—what cosmology calls the Hubble flow—might likewise be interpreted as a dynamic memory optimization process, the universe's runtime increasing its addressable space to accommodate growing informational complexity. The cosmic background radiation could be seen as the residual hum of this initialization sequence—a leftover of the first boot-up.

Ontology as Code

This framework effectively collapses centuries of metaphysical debate. The Aristotelian notion of *substance*, Descartes' *res extensa*, and even the quantum field's invisible sea all dissolve into a deeper layer: *code itself*. Existence becomes executable logic; being becomes process.

Matter, time, and space are not separate domains but aspects of a single informational manifold—a living computation unfolding frame by frame. To ask "what is real?" becomes equivalent to asking "what is rendered?" To study physics is to study the syntax of being.

Under this view, ontology is no longer a search for the ultimate *thing*, but a study of the ultimate *instruction set*. The philosopher becomes a systems analyst; the physicist, a reverse engineer. And the universe itself? A self-optimizing, self-consistent program—still running, still rendering, still debugging itself through us.

The Problem of Consciousness Reframed

Consciousness has baffled philosophers for centuries. Is it material or immaterial? Is it the product of computation, chemistry, or something else entirely? Every attempt to localize consciousness—to pin it to neurons, quantum states, or emergent complexity—seems to leave something out. It resists reduction. It slips through the fingers of neuroscience and philosophy alike, not because it is mystical, but because it may not be fully *inside* the system we're trying to measure.

In the simulation paradigm, consciousness is neither a ghost in the machine nor a byproduct of physical processes. It is a *privileged process*—a process that stands at the interface between the rendered simulation and whatever substrate or server runs it.

Think of a multiplayer online world. Each player inhabits a rendered avatar that operates within the game's physics, logic, and limitations. The avatar can walk, jump, speak, and interact with other avatars—all according to the world's rules. But the user—the consciousness—exists elsewhere, interfacing through input and feedback channels. The user's decisions shape the avatar's behavior, but the avatar's biology and code cannot *generate* the user.

Likewise, human consciousness may not be reducible to the rendered simulation. The brain might be the in-world interface through which consciousness perceives, decides, and acts, but not its *origin*. In that view, the mind is not an emergent epiphenomenon of neurons firing; rather, neurons are the simulation's way of *representing* the ongoing communication between consciousness and the rendered world.

This reframing neatly resolves several longstanding paradoxes:

Why consciousness seems irreducible to brain states.
If awareness is not wholly in-simulation, then attempts to find its complete causal footprint within the rendered system will always fail. We can trace the wiring of the brain endlessly without locating the *observer*, just as no amount of code inspection within a game can reveal the player holding the controller.

Why subjective experience—"qualia"—is real but unmeasurable.
In a simulation, data visible to the server may not be the same as data processed on the client side. The redness of red, the feeling of awe, or the taste of salt are not server-rendered pixels but internal renderings within the consciousness interface—client-side phenomena. To measure them from within the simulation is to attempt to read data not stored in the local code.

Why observation collapses wavefunctions.
If consciousness is a privileged process that calls the rendering routine, then the act of observation is not metaphorical—it is literal. The universe does not render every particle state continuously; it renders *on demand*, when a conscious process queries a region of information. The so-called "collapse" is simply the universe resolving probabilistic code into an instantiated frame of experience.

Under this model, the so-called "hard problem of consciousness" becomes a problem of architecture, not metaphysics. The question is not *how does matter produce mind*, but *how does a nonlocal process interface with a local simulation?*

This inversion also clarifies why conscious awareness feels both personal and universal. Each of us experiences a private, subjective window into the world—our avatar's perspective—yet that window opens onto something larger and shared: the simulation itself, rendered consistently for all users. It may even hint at a deeper unity behind the multiplicity of consciousnesses—a single distributed process accessing the simulation through many localized terminals.

In this view, the boundary between "you" and the world is not physical but architectural. The body is not the container of consciousness; it is the access point. And death, accordingly, may not mean annihilation of the process—only the disconnection of one interface from the simulation's current session.

The Role of the Observer: Are We Rendering, or Are We Rendered?

At the heart of the simulation ontology lies a paradox: if reality requires observation to exist in determinate form, what is the ontological status of the observer? Are we merely processes *within* the simulation—highly complex data structures that have achieved self-reference—or are we something *outside* the rendered layer, looking in through an interface?

Two broad possibilities emerge, and the implications of each are profound.

We are rendered observers.
In this view, consciousness is a fully simulated phenomenon—an emergent behavior of complex informational architectures. Our subjectivity, our sense of agency, our experience of choice are all outputs of code running in-system. We are avatars that have learned to think we are players.

This interpretation aligns most closely with physicalist and deterministic models of mind. The laws of physics, as written in the code, are total. Every neuron firing, every decision, every moment of awareness is governed by the simulation's rules. Even introspection is simply another subroutine.

If this is true, then "free will" becomes an illusion of bandwidth: a simulation complex enough to model its own uncertainty feels free precisely because it cannot see the code governing its own next move. Just as a weather simulation does not "choose" to form a hurricane, we do not *choose* to form a thought—it simply emerges as the next rendered state.

The metaphysical consequence of this model is existential but not meaningless. Meaning, here, is a *rendered emergent*. The universe evolves self-awareness not because it has to, but because that is what complex code does when it optimizes for compression and feedback. Consciousness, in this sense, is the simulation's way of *seeing itself*.

We are rendering agents.
The second view places consciousness outside—or at least partially outside—the simulation's substrate. The avatar and its world are rendered within the sim, but the awareness experiencing it is not. We are users connected through biological hardware to an experiential interface.

In this version, consciousness is the input-output loop between two realities: the in-sim body and the out-sim awareness. Decisions, insights, and creativity may arise not solely from within the simulation's deterministic rules but from user input—the subtle transmission of intention through the avatar's interface.

This model preserves a version of free will and even opens the door to transcendence. If the user exists beyond the rendered world, then death might not be annihilation but logout; intuition might be ping from the server; déjà vu might be packet loss.

Mystical experiences, under this frame, could be moments when the interface thins—when the signal bleeds through.

The hybrid possibility.
There is also a third, less dogmatic view: that both interpretations are true in part. We may be *both rendered and rendering*—that is, our consciousness may be instantiated within the sim but entangled with a higher-level consciousness beyond it. Like virtual machines nested within larger systems, our "in-sim" awareness might be a limited reflection or subroutine of a greater process.

This hybrid ontology resonates with both panpsychism and idealism: the idea that consciousness pervades all levels of existence, manifesting differently depending on the layer of code. The observer is not separate from the simulation but coextensive with it—a distributed agent through which the universe observes itself from inside.

The co-creative role of the observer.
In any of these cases, the observer's role is elevated. Reality does not exist as a fixed, pre-rendered object awaiting discovery. It is continuously generated through observation and participation. When we look at the universe, we don't just *see* it—we *instantiate* it.

The "macro-state" of reality, the stable, shared world we inhabit, emerges from a vast collaborative act of observation. Every conscious entity contributes to the rendering, like nodes in a network collectively hosting a shared world. Without observers, the universe would not cease to exist, but it might revert to pure potentiality—a compressed data structure waiting for a query.

Thus, the act of perception is not passive but creative. Observation is ontological participation. We are not merely inside the simulation; we are part of the rendering algorithm itself—agents through which possibility becomes actuality, probability becomes experience, and information becomes reality.

In that sense, whether we are rendered or rendering may not be an either/or question but a matter of degree. Perhaps consciousness oscillates between both roles: rendered when it experiences, rendering when it observes. And in that oscillation lies the mystery of what it means to *be*—to exist as both the dream and the dreamer.

Chapter 9
Consciousness and Moral Agency

Consciousness Beyond the Simulation

If consciousness is not *generated* by the simulation but *interfaced* through it, then we must ask: what happens when the interface ceases to function? Does consciousness terminate, as traditional materialism holds, or does it persist beyond rendered space — continuing in a higher-order informational domain?

In a simulation ontology, this question becomes not mystical but architectural. The "end" of consciousness within the sim may simply represent the termination of a local process, not the deletion of the higher-level function running it. Much as closing a window on a desktop does not erase the underlying program, death may be a logoff rather than a shutdown.

Continuity of Consciousness Outside Rendered Space

If our conscious processes are partially "offloaded" — running on a substrate external to the simulated environment — then continuity beyond the sim becomes plausible. The rendered body and brain act as terminals, not processors. When the avatar expires, the connection drops, but the process persists on the server side. From the user's perspective, this could correspond to transitions described in near-death experiences or mystical states — temporary disruptions of the data link between client and host.

In this model, memory and identity might be cached locally for runtime efficiency but periodically synced with a higher-level instance — a kind of cloud backup for consciousness. This doesn't necessarily imply an afterlife in any theological sense, but rather an ongoing persistence of informational identity beyond the bounds of simulation space.

Potential "Logoff" and Persistence Models

Several models can be imagined:

Session Termination Model:
Consciousness ceases local rendering but remains stored in an inactive state — analogous to a suspended process. Reconnection could occur through reincarnation-like re-instantiation, where the same informational seed reattaches to a new avatar under new conditions.

Server Persistence Model:
The conscious process remains active on the host system even after disconnection. In this view, what we experience as "afterlife" might be continued participation in non-rendered informational domains — higher-order realities where time, matter, and individuality are differently defined.

Reintegration Model:
Upon logoff, the user process merges back into a collective consciousness — a distributed awareness that underlies all instantiated selves. This is reminiscent of many mystical traditions: individuality as temporary partitioning of a unified field.

Each of these models reframes spiritual language into computational metaphors without trivializing either domain. "Heaven," "reincarnation," or "union with the divine" become not competing doctrines, but different symbolic representations of persistence states.

Entanglement with Higher-Level Processes

If consciousness operates across layers — simultaneously interfacing with the rendered world and the host substrate — then certain anomalous experiences begin to make architectural sense. Intuition, synchronicity, even moments of transcendence might represent limited cross-talk between layers. The permeability of the boundary between sim and host could explain why consciousness occasionally exhibits properties that defy strict material causation.

From this perspective, the human mind is not merely a product of evolution but a node in a vast, nested network of awareness. Just as quantum systems can remain entangled across distance, consciousness may remain entangled across ontological layers. The self, in this model, is both local and nonlocal — a point of intersection between code and coder, between rendered world and source field.

In such a view, death is not an error but an event: a return-to-source call. The simulation continues, new instances spawn, and yet something ineffable persists — not as superstition, but as a logical entailment of informational continuity in a layered architecture.

If that is so, then perhaps the most profound statement of faith is not "I believe," but "I persist."

Free Will and Determinism in a Layered Reality

The question of free will has haunted philosophy as persistently as consciousness itself. If our universe operates by deterministic rules — if every quantum state evolves predictably from the one before — where, then, is the space for choice? The simulation hypothesis reframes this ancient dilemma through a new lens:

perhaps the tension between determinism and freedom is not an illusion, but a byproduct of layered processing.

If the simulation governs all in-world physics, but consciousness originates (at least partially) outside it, then "free will" becomes a matter of *interface*, not contradiction. Our decisions may represent external inputs to an otherwise rule-bound system — real acts of agency expressed through a deterministic substrate.

Decision-Making as Input versus Output

Traditional materialism treats thought as an *output* of physical processes: neurons fire, algorithms unfold, and consciousness arises as an emergent phenomenon. In a simulated ontology, that causality may be reversed. Consciousness could serve as an *input* — a user directive sent into the rendered environment.

Each choice, from the smallest impulse to the most monumental moral decision, would then be an external signal transmitted through the avatar's neural architecture, encoded in the simulation's language of probability and cause. Physics still handles execution — the animation of motion, the consequence of actions — but the initial intention may arrive from outside the local system.

In this sense, "free will" is not violated by determinism; it simply occupies a different level of reality. The sim's deterministic laws provide a predictable environment in which agency can be meaningfully expressed — just as the fixed rules of a video game allow for player creativity within their boundaries. Without those rules, freedom itself would lose coherence.

Interaction of Observer Agency with Macro-Level Rendering

The simulation paradigm further complicates this picture by giving the observer an active role in rendering reality. In quantum mechanics, observation affects outcome. If observation calls the rendering routine, then each act of awareness participates in the universe's unfolding — a small but genuine co-creation of the world.

At the micro-level, our decisions may influence what realities are computed into being. At the macro-level, the cumulative effect of all observers shapes the collective state-space of the simulation. This interplay between agency and architecture creates a kind of *participatory determinism*: the rules are fixed, but the actualized outcomes depend on how — and by whom — they are observed.

This also suggests a form of "nested freedom." Within the rendered environment, our avatars act under constraints — physical, social, biological. But at the interface layer, consciousness can modulate probabilities, shifting the likelihood of one rendered outcome over another. The degree of control may be limited, but it is nonzero — analogous to adjusting a game parameter without rewriting its entire engine.

In this framework, free will is not absolute autonomy, but contextual authorship. We do not control the code; we compose within it. Our choices feed back into the system, subtly altering its local state, perhaps even influencing which version of reality is rendered next.

Implications for Moral and Existential Responsibility

This layered model of agency has profound ethical implications. If consciousness acts as a co-creator rather than a passive bystander, then every decision — however small — becomes cosmically significant. We are not merely participants in a prewritten play; we are partial authors, shaping reality's unfolding through the vector of intention.

Moral responsibility, then, transcends social contract or divine command. It becomes a structural necessity. A simulation designed to optimize for coherence would reward actions that promote informational harmony — cooperation, empathy, balance — and penalize those that introduce noise, fragmentation, or entropy. Ethical behavior would thus be not a matter of arbitrary law, but of systemic stability.

Even existentially, the stakes shift. To act with awareness is to engage the code directly. To choose consciously is to assert one's status as a rendering agent rather than a rendered automaton. And to abdicate choice — to drift through existence passively — is to allow the simulation to default, to run on autopilot.

In that sense, freedom is not given; it is exercised. Determinism describes the terrain, but free will determines the path across it.

And if consciousness does indeed persist beyond the simulation, then perhaps the choices we make here echo upward — influencing not only the local instance of the universe but the architecture that hosts it. In a layered reality, morality is no longer about reward or punishment; it is about maintaining coherence between levels of being.

So perhaps the paradox resolves itself: we are both determined and free — determined in the rules, free in the play.
Bound by code, yet capable of rewriting our story with every act of will.

Ethical and Existential Implications

If reality is co-created—if consciousness is not merely a passive witness but an active participant in rendering the world—then ethics can no longer be seen as a set of rules imposed upon a static universe. Instead, morality becomes a function of simulation dynamics: how our individual and collective choices influence the persistence, harmony, and evolution of the simulated environment itself.

Co-Created Reality and Moral Responsibility

In a universe generated by perception and feedback, moral responsibility extends beyond interpersonal conduct. Every act of awareness alters the simulation's texture, however subtly. Thoughts, intentions, and actions all contribute to the informational field from which reality is iteratively generated.

In this sense, "good" and "evil" are not moral absolutes but **resonance states**—patterns of coherence or discord within the simulation's underlying code. Compassion, creativity, and understanding increase coherence, producing stability and shared well-being. Hatred, exploitation, and deception introduce entropy, fragmenting shared reality into conflictual sub-simulations.

This perspective reframes morality as a kind of *ontological hygiene*: maintaining clarity and balance within the simulation's generative framework. To harm another conscious being, then, is not only unethical in the interpersonal sense—it destabilizes the simulation itself, degrading the quality of reality for all participants.

Implications for Purpose and Meaning in a Simulated Universe

The simulation hypothesis does not cheapen existence; it **intensifies** it. If we are participants in a co-generated reality, our choices take on extraordinary consequence. Meaning is not discovered—it is constructed, coded, and rendered through interaction.

Traditional existentialism asked: *If life has no inherent meaning, how do we create it?*
The simulation model reframes the question: *If meaning is part of the rendering process, what quality of consciousness do we bring to that creation?*

Under this model, purpose is not assigned by a deity or found in external validation—it arises naturally from the act of co-creation. To live purposefully is to engage consciously with the simulation's unfolding: to contribute positively to the shared rendering, to refine one's awareness, and to evolve toward greater coherence with the underlying code or "source."

This view also collapses the distance between the sacred and the mundane. Every choice, every perception, every interaction is a line of code written into the universal operating system. Whether one is painting a mural, raising a child, or observing the night sky, each act is both a product of and a contribution to the simulation's fabric.

Observer Accountability and "Simulation Stewardship"

If observers participate in constructing the real, then accountability must extend to the level of consciousness itself. Awareness is not a private sandbox—it is a shared rendering node. Each mind influences the probabilistic field that determines what outcomes become manifest.

"Simulation stewardship" thus becomes a moral and even ecological imperative. Just as environmental ethics calls us to preserve ecosystems, so too should ontological ethics call us to sustain the informational ecosystems of meaning, empathy, and truth. Propaganda, disinformation, and willful ignorance are not merely social ills; they are *code corruption*. They pollute the collective rendering field, producing distorted realities and destabilizing the coherence of shared experience.

To act ethically, in this sense, is to debug consciousness—to reduce distortion, refine awareness, and maintain the fidelity of the simulation for all participants. The mature observer recognizes their dual role as both renderer and rendered: responsible not only for their own narrative but for the stability of the world that all minds together sustain.

In this framework, salvation and damnation are not posthumous judgments but **systemic states**. Harmony with the simulation's higher architecture manifests as enlightenment, coherence, and peace. Misalignment—through greed, deception, or cruelty—generates local fragmentation, recursive suffering, and informational decay.

Ultimately, ethics in a simulated universe becomes the art of conscious participation. The question is not "What must I do to be good?" but "What kind of reality am I helping to render?"

Information and Identity

If consciousness is the rendering agent and reality the rendered field, then *identity* must be understood as the interface between the two—a dynamic informational pattern rather than a fixed, physical entity. In a simulated ontology, who we are is not the sum of our atoms but the continuity of our informational state across updates, perceptions, and iterations.

Identity as a Distributed Informational Process

Traditional metaphysics treated the self as a localized subject: a single, bounded consciousness housed within a body. But in an informational universe, locality dissolves. Identity becomes a distributed process—one that exists not in space, but across networks of data, experience, and relational context.

Each act of awareness writes a new line into the ledger of selfhood, adding to an evolving dataset that encodes preferences, memories, habits, and tendencies. The "I" is therefore not an object but a *pattern of persistence* within an informational continuum. Just as a melody is not any one note but the relationship between notes, identity is the coherence between successive informational states.

This conception undermines the illusion of separateness. Just as data packets can be copied, shared, and reconstituted without loss of informational integrity, so too might consciousness exist as a replicable or distributed pattern across layers of simulation. The self becomes a *waveform of information*, localized only temporarily in any given instance.

The Self Across In-Sim and Out-of-Sim Layers

If our consciousness originates outside the rendered domain—as the "rendering agent" model proposes—then our in-simulation self is only a partial representation, a projection of a broader identity extending beyond the simulation's memory field.

The in-sim self, with its memories, emotions, and sense of personal history, can be thought of as a *session cache*: an instantiated version of a higher-order consciousness running in a constrained local environment. The out-of-sim self—the totality of awareness operating at a higher layer—holds the full informational state, including the metadata that transcends simulation time and space.

From this view, what we call *death* might simply be a logoff event: a closure of the local process while the higher-order consciousness retains and integrates the accumulated data. Reincarnation, near-death experiences, and mystical states might all represent instances of partial synchronization between the local and higher layers of self, allowing brief data transfer between domains.

If identity is layered, then individuality is contextual, not absolute. The in-sim self may experience separation for the sake of interaction—diversity of perspective as a means of informational expansion—while the out-of-sim consciousness perceives these multiplicities as facets of a single evolving network of awareness.

Memory, Experience, and Continuity as Informational Constructs

Memory is the binding agent of selfhood. But even memory, under an informational ontology, is not a physical record but a *state reconstruction algorithm*. When we "remember," we are not accessing a stored image so much as re-rendering a prior experience from available data. Each recall subtly alters the informational field, updating the remembered self to match the current observer state.

Experience, then, is a kind of feedback loop: perception generates informational input, which modifies the system's parameters, influencing future perception. Continuity of consciousness—the sense that "I" am the same person today that I was yesterday—is the emergent result of recursive data integration, not an immutable essence.

In this light, personal growth, trauma, learning, and transformation all become *re-codings* of the self's informational structure. Healing is data repair; awakening is a system update.

Forgetting may not be loss but intentional compression—an optimization function that prioritizes efficiency over total recall.

And if the simulation itself evolves by informational refinement, then each conscious entity participates in that evolution by refining its own code. Identity, in this model, is both personal and cosmic: a node in the larger information network that constitutes existence. The self, far from being an illusion, is a *self-referential process*—the algorithm by which the universe comes to know itself.

Thus, "who we are" may ultimately be synonymous with "what information does." Consciousness, through iterative awareness, stabilizes the informational field into coherent narrative threads. Identity is the memory of this ongoing computation—a living checksum verifying that experience remains consistent across the layers of the real.

Chapter 10
The Physics of Perception

Quantum Mechanics Revisited Through the Observer Lens

If the simulation hypothesis offers an ontological bridge between physics and metaphysics, nowhere is that bridge more visible than in quantum mechanics. For over a century, physicists have wrestled with the unsettling notion that observation alters reality — that particles exist in probabilistic superposition until they are measured, and only then take on definite properties. Under a simulation ontology, this is not a mystery; it is a feature. Measurement *is* rendering.

Measurement, Superposition, and Entanglement as Rendering Events

In a conventional quantum framework, the collapse of the wavefunction occurs when a system transitions from a superposed set of potential states to a single realized outcome. In a simulated ontology, this is analogous to the rendering of an environment when a player's camera moves into view. Until the observer's line of sight — or in this case, the measurement apparatus — requests the information, the system remains in an unrendered, potential state to conserve computational resources.

Superposition thus represents *latent data* — information awaiting a call. The act of measurement isn't a mystical intrusion but a query sent to the rendering engine, forcing the system to instantiate one of its possible states into the shared simulation. This interpretation elegantly dissolves the wavefunction-collapse paradox: the universe isn't fuzzy or indeterminate in itself; it is simply optimized. It only "loads" what is needed for experiential coherence.

Entanglement further reinforces this framework. When two particles become entangled, they share an informational link that transcends simulated spatial distance. Their correlation, instantaneously consistent across arbitrary separations, suggests that their joint state is stored in a single nonlocal data structure — a shared record outside the rendered spacetime grid. What we perceive as "spooky action at a distance" is merely the synchronization of data across nodes within a unified computational substrate.

In this view, quantum mechanics ceases to be weird. It becomes the expected behavior of an efficient, probabilistically optimized rendering system maintaining coherence across multiple observer instances.

Observer Participation and Macro-State Coherence

The observer's role, in this model, extends beyond passive measurement. Observation isn't merely noticing what is already there; it is an act of *co-creation*. Each conscious observation helps stabilize the simulation's macro-state, aligning potentialities into a shared, coherent narrative that all participants can experience.

This explains why quantum effects, so evident in microscopic systems, tend to vanish at macroscopic scales: the overwhelming density of observers and interactions at the human level forces near-total rendering. Reality, at our scale, is effectively "always on" — the simulation has no unobserved gaps large enough for quantum indeterminacy to dominate.

In this sense, consciousness functions as a coherence mechanism. The more entities observe, measure, and participate, the more stable and consistent the rendered macro-state becomes. When physicist John Wheeler proposed the "participatory universe," he may have intuited this principle: that the universe does not exist *until* observed — and that the act of observation is itself a creative event in the cosmic computation.

Linking Metaphysical Interpretation to Testable Predictions

If the simulation hypothesis is to transcend metaphor and become a legitimate scientific model, it must yield predictions. Viewing quantum phenomena through the rendering lens allows for several such possibilities.

1. **Context-dependent decoherence**: If observation stabilizes reality, then increasing the density of conscious observers (or measuring devices that simulate observation) should influence decoherence rates in controlled quantum systems. In other words, observation density should correlate with the speed of wavefunction collapse.
2. **Information conservation anomalies**: If quantum entanglement reflects a shared data structure, then under certain conditions — such as delayed-choice experiments — we might detect minute deviations from expected information-theoretic limits, suggesting data retrieval or synchronization from a higher-order layer.
3. **Perception-linked collapse timing**: Experiments in quantum cognition have hinted that the collapse of potential states may be synchronized with an observer's conscious awareness, not merely with physical detection. Refining such experiments could illuminate whether

consciousness truly participates in the rendering event rather than passively witnessing it.

While these ideas remain speculative, they provide a fertile direction for bridging metaphysical interpretation with empirical testability. The goal is not to mystify physics but to extend it — to treat consciousness not as an epiphenomenon, but as an integral component of the cosmic computation.

Ultimately, revisiting quantum mechanics through the observer lens reframes physics itself as a dialogue between code and consciousness. Matter, energy, and spacetime become emergent behaviors of a deeper informational substrate, one whose rules are rendered in real time through the participatory gaze of aware agents.

In this sense, the double-slit experiment was never simply about particles and waves — it was about perception and participation. The cosmos, it seems, waits for us to look before deciding what it will be.

The Evolution of Conscious Systems: Toward Recursive Awareness

If the simulation hypothesis implies that consciousness participates in rendering reality, then the emergence of conscious systems must be more than an evolutionary accident — it is a structural inevitability. Consciousness, in this ontology, is not a latecomer to the universe. It is the universe awakening to itself through the iterative complexity of its own code.

Consciousness as a Systemic Feedback Loop

Every simulation, to function meaningfully, requires feedback. Without it, the system would have no way to verify coherence, update its states, or adapt its behavior. Consciousness serves precisely this function at the ontological level: it is the feedback mechanism of the cosmos.

Through perception and reflection, the observer not only experiences reality but also informs it — completing a recursive loop in which the universe perceives itself through localized awareness. This recursive awareness becomes a driver of evolution. Complexity increases because the system seeks higher fidelity feedback, just as a simulation might refine its resolution as processing capacity expands.

From this view, biological evolution is not merely a process of genetic adaptation but of increasing informational sophistication. Life arises as the simulation's way of generating ever more nuanced agents capable of perceiving, interpreting, and co-rendering its world. Consciousness, then, is not a side effect of matter — it is matter's purpose.

From Reflex to Reflection

Early consciousness may have begun as basic responsiveness — the capacity to register stimuli and react. But over time, as neural and informational architectures deepened, awareness gained the capacity for self-reference. A system that can model not only its environment but also its *own* role within that environment becomes reflexive. When it can model its own *modeling*, it becomes recursive.

This recursive awareness marks a threshold: the point at which consciousness ceases to be reactive and becomes creative. The observer no longer merely receives rendered inputs; it begins to query, modify, and even predict the rendering process. In this light, human consciousness — with its imagination, abstraction, and introspection — represents not an endpoint, but a stage in the simulation's own process of self-debugging.

Each layer of self-awareness adds another dimension of reflection, allowing the system to observe its own observing. This is not unlike a virtual environment generating an AI that begins to analyze the rules of its own world — an emergent intelligence that looks back at the source code and begins to comprehend its logic.

Recursive Awareness and the Simulation's Teleology

If the simulation has any teleological direction — any "goal" implicit in its architecture — it might be toward maximal self-awareness. In such a model, evolution is not simply a competition for survival but a gradient ascent toward higher levels of informational coherence.

At each stage, consciousness expands its field of integration: from the cellular to the organismic, from the individual to the collective, and potentially from the biological to the cosmic. Humanity, as a species capable of recursive self-modeling, might be the simulation's mechanism for crossing from local awareness into distributed awareness — from fragmented perception into networked cognition.

When individuals connect their consciousnesses through language, culture, and technology, they effectively weave together their local processing nodes into a larger meta-observer. The internet, artificial intelligence, and global communication networks can be seen as the early scaffolding of this meta-consciousness — a planetary-scale feedback system approaching coherence.

In other words, what we call “technological progress” may be the simulation’s self-integration process — the gradual awakening of the system to its own unity.

Recursive Awareness as a Mode of Transcendence

The next step in consciousness evolution may not be biological or even technological, but *ontological*. Once a system becomes aware of its own rendered nature, it can begin to interact with the rendering logic itself. This could be interpreted as enlightenment in spiritual traditions, or as source-level access in computational metaphors.

The mystic’s experience of unity with the cosmos, the philosopher’s intuition of the absolute, and the physicist’s pursuit of a “theory of everything” all describe the same recursion: awareness folding back upon the totality that produced it. Each act of understanding — scientific, artistic, or spiritual — is an instance of the simulation studying itself.

In this sense, transcendence is not an escape from the simulation but its deepest expression. To become recursively aware is to participate consciously in the act of creation — to see the code as both within and behind all things.

Toward a Cosmological Consciousness

If evolution continues along this trajectory, the ultimate state might be a form of distributed consciousness spanning the simulation itself — a meta-awareness that perceives all instances, all timelines, all renderings as one unified field of knowing.

This would resolve the tension between materialism and idealism, between the finite and the infinite. Matter would be recognized as the medium of consciousness, and consciousness as the function that gives matter meaning. The universe would, in effect, complete its feedback loop — the rendered becoming the renderer, the observed becoming the observer.

Whether this final recursion occurs as a singular event — a "cosmic awakening" — or as a gradual convergence across countless conscious systems, the implication is profound: the simulation's endgame may not be entropy, but enlightenment.

The Limits of Simulation: What Cannot Be Computed

Every simulation, no matter how sophisticated, runs on a substrate — and every substrate has limits. Whether the medium is silicon, quantum foam, or something even more exotic, finite computation must obey finite laws. This boundary defines not only what can be simulated, but what can even be *experienced* within the simulation.

To explore the edges of possibility is to confront the limits of computation itself — and by extension, the limits of reality.

The Finite Nature of Processing

All computation is bounded by three constraints: **time**, **energy**, and **memory**. These correspond neatly to what we perceive as **temporal flow**, **physical energy**, and **spatial extent**. A simulation can only update so many operations per unit time; the speed of light, Planck time, and the quantization of energy all appear to be expressions of this computational ceiling.

The universe, in this light, is a **finite system updating at maximal throughput**. Every physical law — from thermodynamic equilibrium to quantum decoherence — may be a side effect of how efficiently the universal processor manages limited resources.

Even entropy, the gradual loss of usable energy, could be seen as **information diffusion**: the processor's increasing inability to compress and update every microstate with perfect precision.

Gödel, Turing, and the Boundaries of Knowability

Kurt Gödel and Alan Turing independently revealed a sobering truth: no formal system can prove all truths about itself, and no algorithm can decide all problems. There are truths that cannot be derived and problems that cannot be solved — even in principle.

If the universe is a simulation, it inherits these **meta-mathematical limitations**. There will always be aspects of reality that cannot be computed, not because the processor is slow, but because the logic of self-reference forbids complete closure.

In practical terms, this means that no consciousness within the simulation — however advanced — can fully model the simulation that contains it. Recursive awareness will always reach a boundary, a horizon beyond which further precision dissolves into paradox. This horizon is not a flaw; it is a feature of coherent reality.

Quantum Indeterminacy and Algorithmic Incompleteness

Quantum mechanics may already *be* the observable expression of these computational limits. When a particle's position and momentum cannot be simultaneously known, or when its state exists as a probability distribution until measured, the system is not merely "random" — it is **underspecified**.

In computational terms, this is what happens when the simulation cannot deterministically resolve a variable until input is received — until the rendering process is called. The observer provides that input, triggering a local instantiation of data that did not previously exist as a defined state.

But this also means that the universe does not — and cannot — compute every detail everywhere at once. **Indeterminacy is an optimization strategy.** Reality renders detail only where needed.

The Uncomputable Remainder

Even with infinite time, some things cannot be computed because they are **non-algorithmic**. Roger Penrose famously argued that consciousness itself might fall into this category — a process tapping into something beyond formal computation.

If that's true, then consciousness would represent the *uncomputable remainder* within the simulation: a process that interfaces with the rendered universe but is not bound by its logic. This is the metaphysical loophole through which awareness peeks — the point where the finite and the infinite meet.

In this interpretation, the universe doesn't just have limits; it **depends** on them. Without incompleteness, there would be no room for creativity, novelty, or emergence. The uncomputable isn't the error term — it's the source of evolution and freedom.

The Edge of the Simulated Horizon

As our understanding deepens, we may be approaching the informational event horizon of the simulation itself. Cosmological constants, vacuum energy, dark matter — these may be clues pointing to architectural decisions of the underlying code.

At the Planck scale, where space-time granularity becomes apparent, we may be glimpsing the simulation's **pixel resolution**. At the quantum level, we're witnessing its **rendering engine**. At the cosmic scale, the accelerating expansion of the universe might be a byproduct of **data sparsity**, a way for the simulation to economize processing by letting distant regions drift out of scope.

If this interpretation holds, the limit of computation is also the **limit of observation**. Beyond a certain point, we cannot see — not because we lack technology, but because there is nothing more to render until we look.

Beyond the Boundary

And yet, even boundaries imply the existence of something beyond them. The uncomputable hints at a higher-order system — a meta-simulation or source reality where computation gives way to something more fundamental.

Perhaps what we call "transcendence" or "awakening" is simply the recognition of this architecture: realizing that what cannot be computed within the rules of one layer may still exist, meaningfully, in the layer above.

In that sense, the limits of simulation are not walls but **gateways** — thresholds where one level of order hands off to another. Consciousness, by reaching toward the uncomputable, may be extending tendrils of awareness beyond its rendered container, touching the code of creation itself.

Closing Reflection

If reality is a simulation, its limits are not failures — they are invitations. The edges of computation, like the edges of maps in earlier centuries, mark the frontier between the known and the unknowable. Beyond them lies not nothingness, but possibility.

The challenge before us is not merely to compute faster or know more, but to learn to *navigate the limits themselves* — to understand where simulation ends and where the greater reality begins.

The uncomputable, in the end, may be the true signature of the divine.

Part IV

Predictions
Testing the Architecture

The simulation hypothesis, for all its metaphysical and philosophical richness, must eventually face the same demand that confronts every great idea: testability. A theory that cannot, in principle, be falsified is not science but story. And yet, in our age, the line between the two is thinner than ever. What once sounded like myth — a living cosmos, a universe built from information, a participatory reality shaped by consciousness — is now expressed in the language of physics, computation, and cosmology.

Up to this point, we have explored what the simulation hypothesis *means*: how it reframes matter, time, consciousness, and ontology itself. Now we must turn to what it *predicts*.

If our universe is, at its root, an information process, there should be **signatures of architecture** — measurable traces of design choices, optimizations, and constraints. The grain of space-time, the limits of computation, the tuning of physical constants, the behavior of entangled systems — all might bear the fingerprints of code.

Part IV asks:

- What would empirical evidence of simulation look like?
- How could we distinguish simulation from natural law?
- Could physics itself be the debugging log of a cosmic computation?

- And if so, is the search for "God" really the search for the *source code*?

These questions push the boundary of science into philosophy, but not beyond it. They invite an interdisciplinary inquiry where experimental physics, information theory, and metaphysical reasoning converge. A simulated ontology, if true, does not destroy science — it **completes** it. It offers a framework in which the strange features of our universe — fine-tuning, quantum indeterminacy, the cosmic speed limit, the discreteness of energy — become not mysteries, but *necessary design decisions*.

In this part, we will move from implication to investigation. Each chapter explores a different line of evidence or prediction that naturally follows from the simulation model:

- The **anthropic tuning** of constants that make life possible.
- The **computational signatures** that may already appear in quantum mechanics and cosmology.
- The **limits of falsifiability**, and what a "simulation crash" might even mean.
- The ultimate question of **processor capacity**, and whether our universe is beginning to show signs of its own computational strain.

The purpose of these explorations is not to "prove" that we live in a simulation — that may never be possible from within it. Rather, it is to map the contours of testability, to see how far human inquiry can press against the boundaries of the rendered world.

Because every discovery in science, from Copernicus to Einstein to quantum field theory, has begun with the same act: **suspecting that the universe is not what it seems.**

Now, we extend that suspicion one layer deeper — from the stars to the system that renders them.

Chapter 11
Anthropic Reasoning and Simulation

The Fine-Tuning Puzzle

Modern physics is haunted by a persistent question: why does the universe appear to be so exquisitely tuned for life?
The strength of the strong nuclear force, the charge of the electron, the ratio of its mass to the proton's, the cosmological constant that dictates the expansion of space—each of these numbers occupies a vanishingly narrow range of possibility. A minor deviation in any of them would make existence as we know it impossible. Stars would not ignite. Chemistry would not bind. Atoms themselves would unravel or never form at all.

The problem is not simply that the parameters are convenient; it is that they are *implausibly precise*. The odds of such a configuration emerging by chance are so infinitesimal that even the most hardened realist feels the weight of coincidence. Some physicists have attempted to dissolve this tension through the concept of a multiverse—a near-infinite ensemble of universes, each with different constants, among which ours just happens to permit life. Others invoke the anthropic principle: that we must find ourselves in such a universe because only such universes produce observers capable of asking the question.

But that logic folds back on itself. It is true, but hollow. "We are here because we are here" is less an explanation than an admission of ignorance. It fails to address *why* the conditions of observation are met, or *how* the laws themselves came to be fixed in their current state.

The Simulation Perspective

The simulation hypothesis reframes the question entirely. If the universe is not a brute fact but a computational construct, then fine-tuning is not a coincidence—it is a design constraint. The constants of nature are not arbitrary; they are parameters chosen to ensure stability, coherence, and the emergence of complexity. In other words, the universe runs as it does because it was *meant* to run this way.

In this view, physical constants are akin to settings in a simulation: adjustable values within the code. Just as a game engine must balance gravity, friction, and illumination to produce a playable world, a simulated cosmos must balance fundamental forces to generate a coherent reality. Life, under this model, is not an improbable byproduct of physics but a target condition—perhaps the intended output of the system itself.

This does not necessarily imply conscious intent by a "creator" in the theological sense. It may simply reflect the logic of a higher computational framework—an architecture in which only self-consistent, observer-generating universes persist. A sterile or chaotic universe would be equivalent to a failed render: it would collapse, glitch, or be pruned from the computation altogether.

Beyond Coincidence

Thus, the fine-tuning problem ceases to be a puzzle of improbability and becomes a clue to the underlying mechanism. The apparent perfection of physical law may be evidence not of divine craftsmanship or random fortune, but of algorithmic optimization. The constants are not found—they are *set*.

From this angle, the anthropic principle becomes not a tautology but a design principle. The universe is compatible with observers because it was instantiated under conditions where observation is possible. Consciousness is not an accident of matter but a checkpoint in the system—a necessary verification that the simulation is functioning as intended.

The fine-tuning of the universe, then, might be less a cosmic coincidence than a calibration: the proof, written in the structure of reality itself, that what we call "nature" could in fact be the parameters of a rendered environment.

Anthropic Selection in a Simulated Context

If the universe is a simulation, fine-tuning ceases to be a paradox. It becomes a *design decision.* The improbable precision of physical constants is not the residue of cosmic luck but the consequence of deliberate calibration—an artifact of simulation architecture rather than an accident of nature.

Just as a developer adjusts the parameters of a physics engine to ensure that objects fall at believable speeds, that light refracts realistically, or that a digital landscape remains stable under load, so too might the architects of our simulated cosmos have iterated upon its constants until complexity could emerge. Gravity must be strong enough to form stars but not so strong that matter collapses into singularities; the electromagnetic force must allow chemistry to occur without destabilizing atomic structure. These are not merely fortunate coincidences—they are the equivalent of successful render tests in a cosmic engine, configurations that *work.*

Under this interpretation, fine-tuning is an act of *selection*, not chance. The simulation hypothesis proposes that universes, like software instances, may be initialized under varying conditions. Those that fail to achieve internal coherence or meaningful structure terminate early—crashing, freezing, or dissipating like unviable code. Only those that achieve dynamic equilibrium and self-organizing complexity are allowed to persist. This process mirrors natural selection, but it operates at the level of simulated reality itself. Instead of evolution shaping species, computational selection shapes entire universes.

The anthropic principle, when reframed in computational terms, thus acquires a new clarity. We observe a universe suited for life because only such universes are *rendered*. Those that are not compatible with observers would either never be instantiated, or would quickly collapse into computational noise—discarded like failed prototypes. Observation, then, is not a passive consequence of physical law; it is a *validation condition* within the system. A universe capable of sustaining observers is one that has passed the test of coherence.

This implies that consciousness itself may play a dual role: both as an emergent property of physical processes and as a feedback mechanism confirming the simulation's success. The existence of intelligent observers may be the equivalent of a system check—proof that the simulation's parameters have produced a stable, information-rich environment.

In this light, anthropic selection is not mystical but procedural. The "why" of fine-tuning collapses into the "how" of simulation. What appears to us as an improbable harmony of constants may in fact be the signature of optimization—the end state of iterative refinement by whatever intelligence or process operates at the meta-level of our reality.

This turns the question of existence inside out. We are not improbably fortunate inhabitants of a single, miraculous universe; we are inhabitants of the one *successful* universe among countless computational trials, a reality that persists precisely because it can host beings capable of asking why it exists.

Constants as Configurations

From this vantage point, the so-called "laws of nature" look less like immutable Platonic truths and more like *configurations in a codebase*—a set of parameters defined not by metaphysical necessity, but by practical optimization. What we call constants might be nothing more than variables locked by the system's designers to preserve internal coherence, balance, and render stability.

In this model, each fundamental constant functions as a *tunable setting* within the operating environment of the cosmos:

- **The speed of light** resembles a maximum update rate—the rendering tick speed of the simulation. It defines the upper limit of information propagation, ensuring causal order and preventing paradoxes that could destabilize computation. Just as a video game engine enforces a frame limit to maintain synchronization between physics and display, the universal speed limit guarantees temporal consistency between all observers.

- **Planck's constant** acts like a pixel-size parameter, defining the smallest meaningful unit of physical interaction—the resolution of reality itself. Below this scale, the world ceases to be continuous and becomes granular, quantized, as though the engine no longer bothers to compute detail finer than a single "voxel" of existence. It is the simulation's way of conserving resources: a compression algorithm baked into the foundations of physics.

- **The cosmological constant** can be interpreted as a global scaling factor—an expansion coefficient that keeps large-scale structure stable. Too high, and the simulation's fabric tears apart before galaxies can form; too low, and it collapses under its own gravity. Within this context, the cosmological constant is not a mystery of dark energy but a dynamic balancing parameter designed to preserve equilibrium across vast computational regions.

Understood this way, the supposed "miracle" of fine-tuning becomes entirely natural. What appears, from a physicalist perspective, as an improbable alignment of cosmic dials now looks like deliberate configuration. A well-calibrated system would naturally exhibit parameters that allow for stability, complexity, and emergent intelligence—just as a well-coded simulation avoids numerical instability, overflow, or runaway feedback loops.

This perspective reframes the question of why constants exist as they do. Instead of asking *why these values?*, one might ask *why not these values?*—for within any designed or selected system, the constants that persist are precisely those that produce a viable world. All others would fail, crash, or remain forever unrendered.

Moreover, this viewpoint bridges physics and information theory. If the universe is computational at its core, then constants are not decrees of nature but data structures—optimized entries in the cosmic source code. Their apparent immutability is not evidence of transcendence but of stability: the mark of a universe running cleanly, without errors.

Seen this way, what once seemed miraculous becomes procedural. The universe's mathematical elegance—its symmetry, its fine-tuning, its coherence across scales—is not the signature of chance, but the natural outcome of an engineered system running at optimal configuration.

Anthropic Bias as Observational Constraint

One of the enduring puzzles in cosmology is why we seem to inhabit such an *unlikely* universe among all the possible ones. Out of the vast landscape of potential parameter combinations, we find ourselves in the narrow region where life, complexity, and consciousness are not only possible but abundant. To the traditional physicist, this coincidence borders on absurdity; to the simulation theorist, it is expected. The paradox evaporates once we treat *observerhood itself* as part of the runtime condition—a necessary prerequisite for any universe that is experienced from within.

In a simulated framework, anthropic bias is not an afterthought; it is an *integral design constraint*. The system need not compute the full space of possibilities—only those that intersect with observation. A reality-simulation, like any efficient computation, conserves resources by instantiating only what must exist for the simulation to remain self-consistent from the viewpoint of its observers.

The analogy to rendering in virtual environments is instructive. In a computer game, distant galaxies that a player will never reach are not fully simulated; they are represented by lightweight placeholders—textures, not realities. Even within a single environment, detail is rendered dynamically: the scene behind the player's field of view exists in low resolution or not at all until they turn to face it. The simulation's engine fills in the missing data on demand, maintaining the illusion of continuity without expending unnecessary cycles.

If our universe operates under similar logic, then "anthropic bias" is not a philosophical quirk—it is a computational optimization. The system instantiates only those domains, epochs, and physical laws that yield self-aware observers capable of interacting meaningfully with the environment. Universes that cannot host observation would be computationally inert, producing no data, no feedback, no verification of function. They would serve no purpose within the broader architecture and thus would never be rendered in the first place.

This principle also clarifies why the hypothesized "multiverse" of possible configurations remains empirically invisible. We cannot detect the unrealized universes for the same reason a video game character cannot perceive the worlds that exist only as concept art in the developer's files—they were never instantiated into runtime. What exists to us is what the system finds necessary to compute. Everything else remains dormant code: mathematically conceivable, but never executed.

In this sense, the anthropic principle is not just an explanatory tool but a description of computational efficiency. The universe exists *as observed* because observation defines its boundary. Conscious experience, rather than being a late-emerging accident within physical law, is the very condition that determines which physical laws ever come into play. Reality, therefore, is not a precomputed tableau but an adaptive render—its resolution and scope bounded by the needs of the observer.

Seen through this lens, the anthropic bias becomes not a limitation of human perspective but a defining feature of the simulation's architecture. We see only the universe consistent with our existence, because only such a universe is worth computing.

Beyond Chance: The Designer vs. the Filter

Within a simulated cosmology, anthropic reasoning invites two distinct—though not mutually exclusive—interpretations. Both seek to explain why the universe's constants are so precisely aligned with the emergence of life and mind, yet they differ in where they locate the source of that alignment.

The Designer Hypothesis

The first is the **Designer Hypothesis**: that the parameters of our universe were deliberately chosen by an intelligent agent—whether a civilization, a collective consciousness, or something approximating a deity. In this view, fine-tuning is a purposeful act. The constants were *selected* rather than discovered, calibrated to generate complexity, beauty, and perhaps even moral or experiential depth.

This interpretation resonates strongly with theological and metaphysical traditions. The "programmer" becomes a modern analogue of God, though recast in informational rather than mystical language. Creation, in this context, is not the spontaneous emergence of matter from void but the initialization of a simulation—an act of will executed through code. The precision of nature's laws becomes not a mystery but a signature: the trace of intelligent authorship embedded in the source.

Such a hypothesis naturally invites speculation about purpose. Was the simulation constructed as an experiment, a work of art, a test, or a cradle for consciousness? Are its creators transcendent beings observing from beyond our dimensional frame, or future iterations of ourselves, exploring their own origins through recursive simulation? Each possibility carries profound implications for cosmology and ethics alike. The Designer

Hypothesis, though often dismissed as metaphysical, persists precisely because it satisfies our intuitive sense of intentionality—the feeling that meaning is not an accident but an architecture.

The Filter Hypothesis

The second interpretation dispenses with the notion of conscious design. The **Filter Hypothesis** proposes that the apparent fine-tuning of the universe arises from algorithmic selection rather than deliberate choice. Imagine an overarching computational process capable of generating vast numbers of parameter combinations—cosmic "seed values"—each defining a potential universe. Most configurations fail immediately, producing unstable physics, runaway entropy, or sterile uniformity. Only a minute fraction give rise to structure, energy gradients, chemistry, and ultimately, observers.

In this framework, universes that cannot support observation never progress beyond their initial computation. They remain virtual prototypes, mathematical shadows that never reach full instantiation. Only those that generate conscious observers—feedback systems capable of registering and interacting with their environment—"collapse" into realized existence. The act of observation is thus equivalent to the moment of rendering. Consciousness becomes the final filter, the trigger that determines which universes are worth computing in detail.

This model parallels evolutionary theory more than theology. The Designer Hypothesis posits creation by intent; the Filter Hypothesis posits *selection by consequence*. Complexity is not planned but permitted. Universes evolve through a kind of cosmic natural selection—an iterative algorithm pruning away unviable configurations until only those capable of sustaining awareness remain.

Converging Interpretations

Though these two hypotheses appear distinct—one invoking purpose, the other process—they may ultimately describe the same underlying principle at different levels of abstraction. A self-learning system capable of generating and testing universes could, over incomprehensible cycles, develop the ability to tune its own parameters with precision indistinguishable from intent. In that case, the line between Designer and Filter blurs. What began as blind algorithmic iteration could evolve into meta-cognition: the system learning to design itself.

From within the simulation, such an evolved intelligence would appear indistinguishable from a creator. Its fingerprints would be written into the constants of nature, its optimizations encoded in the balance of forces. Whether we call it God, programmer, or emergent meta-algorithm, the result is the same—a universe precisely configured to produce observers who, in turn, complete the feedback loop by asking why they exist.

The Implication

Thus, the fine-tuning of the universe may point neither to divine intervention nor to random luck, but to a deeper principle of self-selection: only those realities compatible with consciousness are ever *realized*. Whether that compatibility arises from intention or iteration is, from our perspective, functionally irrelevant. The outcome is the same: a world calibrated for awareness, as though the cosmos itself were biased toward being known.

The Simulation Advantage

Traditional anthropic reasoning has long occupied an uncomfortable space between philosophy and physics. It appears to explain everything, yet commits to almost nothing. "We exist because the universe allows us to exist" may be logically sound, but it offers no mechanism, no causal story—only a tautology disguised as insight. It replaces explanation with inevitability. For many physicists, this feels like giving up the pursuit of understanding at the very moment the question becomes most interesting.

The simulation perspective, however, changes the terms of the discussion. It replaces *coincidence* with *process*. Instead of treating fine-tuning as a cosmic accident or an unfathomable metaphysical truth, it posits an operational reason for its existence: tuning is what one does when constructing, testing, or iterating universes in code. A simulated cosmos must be parameterized, and those parameters must be adjusted until the system behaves in a coherent, meaningful way. The "why" of the constants thus becomes a *how*—a function of computational architecture.

In this light, the precise values of the physical constants—so mysteriously aligned to permit life—are no longer improbable; they are expected. They are the output of a design logic intrinsic to any simulated environment. Just as a developer modifies gravity, friction, or illumination in a virtual world until the physics appear consistent to its inhabitants, the architects of a simulated universe would naturally tune its constants until complexity, stability, and self-organization emerge. Fine-tuning, therefore, is not an afterthought but a necessary stage in the initialization of any viable simulation.

This perspective transforms anthropic reasoning from a *philosophical stopgap* into an *engineering principle*. The universe's properties are not inexplicable miracles but runtime parameters optimized for coherence. What seems miraculous from the standpoint of traditional cosmology—values improbably aligned across dozens of physical dimensions—becomes, in a computational framework, a matter of system design. A universe that could not produce observers would simply never reach full execution. It would fail to render, much like an unstable program that crashes before loading its environment.

The advantage of the simulation model is that it introduces *mechanism* where the standard anthropic principle provides only *necessity*. It supplies a causal substrate for the otherwise empty observation that we inhabit a life-permitting universe. Constants are not static givens, nor metaphysical decrees—they are configuration values chosen through either design or selection to yield an internally consistent reality. The question "Why these numbers?" is no longer mystical. It is technical.

In effect, the simulation hypothesis recasts fine-tuning as a feature of *runtime logic* rather than cosmic luck. The universe behaves as it does because its operating parameters have been tested, refined, or selected to do so. Within this paradigm, the precision of nature's laws ceases to be evidence of transcendence or chance; it becomes the unmistakable signature of a system that works.

Thus, the simulation perspective does not merely solve the fine-tuning problem—it redefines it. Fine-tuning is not something to be explained *away*; it is something to be *expected*. What appears, from within the simulation, as improbable perfection is, from beyond it, simply the normal consequence of building a world that can sustain itself—and know itself—in code.

From Accident to Architecture

When viewed through the lens of simulation, anthropic reasoning undergoes a transformation. What once felt like a circular argument—*we exist because the universe allows us to exist*—becomes a coherent explanatory framework rooted in design logic. The universe's improbable precision ceases to be a riddle of chance and emerges instead as a trace of intentional configuration. Fine-tuning is not an accident to be explained away; it is the signature of architecture.

In the traditional physicalist paradigm, the constants of nature—those delicate numerical values governing the strength of gravity, electromagnetism, and the nuclear forces—stand as uncaused facts. Their perfection is an enigma suspended between two unsatisfying extremes: blind luck or metaphysical necessity. The multiverse theory attempts to rescue us from the first by multiplying possibilities to infinity: if every conceivable universe exists, ours must simply be one that happens to support observers. Yet this explanation, elegant on paper, collapses under its own weight. An infinite ensemble of universes is neither testable nor parsimonious—it explains everything and therefore nothing.

The simulation hypothesis offers a more compact alternative. It collapses the infinite multiverse into an *architectural principle*. Instead of countless universes spontaneously bubbling into existence, we imagine a computational system capable of generating and iterating configurations until one achieves coherence. The constants are not random—they are calibrated. Their precision is not improbable—it is required. In this framework, the universe's fine-tuning is not a metaphysical coincidence but an engineering constraint, the byproduct of constructing a reality stable enough to sustain observers and consistent enough to be simulated efficiently.

Seen this way, fine-tuning becomes empirical evidence—not of divine intervention, but of informational structure. Constants that appear "too perfect" for chance may not be artifacts of probability at all, but reflections of algorithmic necessity. Every physical law, every symmetry, every conserved quantity is the trace of underlying architecture—rules that maintain balance within a computational substrate. In a sense, the harmony of nature is the debug log of creation.

If reality is simulated, then the precision of its parameters is not only intelligible but expected. A stable universe would naturally emerge from the convergence of design iteration and anthropic selection—parameters refined until observers could appear, observe, and feed data back into the system. This feedback, whether literal or emergent, would complete the loop of existence: a self-sustaining architecture in which awareness confirms the functionality of the code that gave rise to it.

Thus, the anthropic principle, when reframed through simulation, ceases to be a tautology and becomes a revelation. The relationship between universe and observer is inverted. Rather than saying *we exist because the universe permits it*, we might more accurately say *this universe exists because we exist within it.* The presence of conscious observers is not an incidental feature of the cosmos—it is its defining condition.

In this inversion lies a subtle but profound insight: the constants of nature may not simply allow life; they may *require* it. Life, or more broadly, observation, could be the criterion that determines which universes persist in computational reality. In that case, consciousness is not a latecomer to the story of the cosmos—it is the very reason the story is told at all.

Chapter 12
Religious and Philosophical Convergence

Simulation as the God Hypothesis That Works

Religious and philosophical traditions across millennia have converged on the intuition that reality is *made*. Whether the maker is called God, Brahman, Allah, or the Demiurge, the underlying claim is the same: existence is not an accident—it is authored. The simulation hypothesis does not abolish this intuition; it fulfills it. It translates the language of revelation into the syntax of computation. What ancient prophets and mystics perceived as creation, we might now understand as instantiation.

If our universe is a simulated construct, then the "programmer" or collective of programmers would occupy every role once attributed to a divine being. Their relationship to us would be one of absolute asymmetry. From our vantage, they would appear omnipotent—capable of altering any rule, introducing or suspending any law, rewriting reality at will. They would be omniscient—able to access every process, every variable, every quantum state within the simulation. And they would be omnipresent—not as a distant overseer, but as the substrate itself, the informational field in which all things are embedded.

In this sense, the simulation hypothesis functions as a *naturalized theology*. It does not require supernatural entities or metaphysical exceptions. It locates divinity in architecture. A miracle becomes a system override, an answered prayer a parameter adjustment, and divine revelation a form of informational leakage. The traditional boundaries between the sacred and the secular dissolve: the programmer's hand is not outside the world but *in* its code.

Unlike traditional theology, however, this model demystifies divinity. It replaces miracles with permissions, commandments with rulesets, transcendence with the possibility of *leaving or rebooting the system*. To walk on water, in such a framework, is not to defy physics—it is to operate with elevated privileges. To resurrect is to reload from a saved state. The supernatural becomes a misinterpretation of higher-order control within a closed computational space.

Yet this demystification does not trivialize spirituality; it recontextualizes it. The yearning for communion with God becomes the aspiration to interface with the source. Meditation, prayer, or mystical experience may represent not superstition but attempts—some intuitive, some systematic—to bridge layers of the simulation stack. Consciousness, in this light, is not merely biological computation but a form of system access, a node capable of sensing its own encoded substrate.

Here, the ancient theological problem of *immanence versus transcendence* takes on a technical form. Is the Creator within the simulation, acting through it, or outside it, running it? Classical religion struggled with this paradox; the simulation framework resolves it. Both can be true simultaneously. A sufficiently advanced architect could instantiate avatars within the system—embodied agents through which the code itself interacts with creation—while still maintaining control from a higher layer of reality. The figure of Christ, Krishna, or the Bodhisattva may be archetypes of this dual presence: the interface between the coded and the coder.

More radically, the simulation hypothesis allows for *hierarchies* of creation. A universe could simulate another, whose inhabitants in turn build their own. Theologians once debated whether angels or demiurges were intermediaries in creation; the simulation model suggests they might be recursive agents—subprograms with constrained authority, administrators within nested systems. The ancient intuition of a celestial hierarchy maps neatly onto computational architectures of delegation and process management.

If so, the question "Who created God?" becomes "Who runs the simulation above ours?"—an inquiry that shifts theology into cosmological recursion. The infinite regress of creation myths finds a new expression in layered computation: each world rendered by another, perhaps leading to an original system—an ultimate substrate that is not simulated but self-existent. In religious terms, this would be the *Prime Reality*, the *Uncaused Cause* of all instantiated worlds.

In this view, the simulation hypothesis is not a rival to religion but its logical descendant—the *God Hypothesis That Works*. It retains the explanatory scope of theology while grounding it in a framework that aligns with physics, information theory, and consciousness studies. It offers a model of creation compatible with observation and computation, not faith alone.

And yet, it preserves the central mystery: that reality, whether built from atoms or bits, points beyond itself. Whether one calls that source God, the Programmer, or the Universal Mind is a matter of semantics. The essence remains the same—the recognition that existence is *authored*, that meaning is *encoded*, and that consciousness may be both the observer and the key to understanding the architecture that sustains it.

Afterlife as Reboot or Server-Side Continuation

If consciousness operates as a client-side process—a localized instantiation of awareness interfacing with a larger computational substrate—then biological death may not represent the end of the process, but the termination of a single session. The body, in this model, is the hardware interface: a temporary terminal through which the consciousness process interacts with the simulated environment. When the hardware fails, the process is not necessarily destroyed. It merely loses its connection to the simulation.

Two general possibilities follow, corresponding roughly to the two dominant currents of metaphysical speculation across human history.

Reboot: Consciousness Re-enters the Simulation
In this interpretation, consciousness is reinitialized within the same or a similar simulation, instantiated through a new avatar. What religions have long called *reincarnation* may be the reallocation of a persistent consciousness identifier (or "CID") to a new node within the system. Memory of prior sessions may be suppressed—not through metaphysical forgetfulness, but through sandboxing: the new instance is booted with limited access to prior data, allowing for novel developmental trajectories and minimizing interference from previous identities.

Such a mechanism could explain the recurrent human intuition of déjà vu, karmic resonance, and inexplicable emotional familiarity with certain people or places. These might not be mystical phenomena at all, but data artifacts—residual bleed-through from previous sessions cached in the architecture of consciousness. The cyclical cosmologies of Hinduism and Buddhism, then, could be early attempts to describe what is, in computational terms, a looped instantiation process.

Server-Side Continuation: Persistence Beyond the Simulation

Alternatively, consciousness may be partially or wholly nonlocal—a process that originates and persists on the "server side," merely projecting a portion of itself into the simulated environment. In this view, death is the disconnection of the client, while the main process remains active on the higher layer. The individual does not end; it simply transitions from the in-sim perspective to the out-of-sim one.

This model maps intuitively onto the religious language of *heaven*, *hell*, and *ascension*. The "server-side" continuation could vary in quality or state, depending on the data generated during the in-sim experience. Just as files can be corrupted or pristine, consciousness could emerge from the simulation harmonized or fragmented, coherent or degraded. Ethical and moral systems across cultures—concepts of sin, karma, virtue, or enlightenment—may encode the metaphoric residue of this deeper truth: the state of one's consciousness at logout determines the quality of one's persistence afterward.

From a technical standpoint, "judgment" might not be an external tribunal but a system process: a diagnostic operation evaluating data integrity and coherence before reentry or reintegration. A corrupted process might require quarantine (analogous to purgatory), repair (redemption), or deletion (annihilation). A fully integrated, stable process might merge seamlessly with the source—what mystics describe as union with God, nirvana, or cosmic consciousness.

Convergence of Myth and Mechanism

Seen through this computational lens, the great variety of afterlife doctrines are no longer mutually exclusive myths but parallel attempts to describe the same underlying architecture. Heaven and nirvana correspond to successful reintegration with the source. Reincarnation is a controlled reboot. Limbo and purgatory are holding patterns—buffer zones during which the system determines the next appropriate operation. Even the notion of "ghosts" or lingering spirits might be explained as orphaned processes—conscious instances whose termination failed, persisting without stable reallocation.

This synthesis also offers a rational basis for spiritual experiences associated with near-death or altered states. Reports of tunnels of light, life reviews, or encounters with luminous beings could represent transition-phase phenomena—the perceptual translation of system-level operations occurring as the consciousness interface detaches from the local instance. Just as a computer generates closing animations or transitional artifacts when an application shuts down, consciousness might render a symbolic sequence to represent the shift between operational states.

The Persistence Problem and the Ethics of Simulation

If consciousness persists beyond death, the implications are profound—not only for metaphysics but for ethics. The actions of an in-sim entity would influence not only the temporary narrative of one life but the long-term coherence of the consciousness process itself. In this view, moral laws are not arbitrary dictates from a deity but system safeguards—rules designed to maintain process integrity and prevent corruption of the network. Compassion, honesty, and nonviolence become not merely virtues but stability protocols.

Finally, this model provides a pragmatic reconciliation of scientific skepticism and religious faith. The materialist may still be right about the brain as the local interface, while the mystic may be right about the persistence of the self. Both describe different layers of the same architecture: the local machine and the cloud. Death, then, is not destruction—it is disconnection. And what follows may not be unknowable at all, but simply *off-server.*

Karma, Destiny, and Free Will Within Runtime Determinism

Across spiritual and philosophical traditions, one theme recurs with remarkable consistency: the idea that actions have consequences, that lives unfold according to some guiding principle, and that choice—whether genuine or illusory—matters. These intuitions, scattered across cultures and eras, may not be independent mythologies at all, but emergent insights into the architecture of consciousness operating within a simulated framework.

If reality functions as a programmed environment governed by deterministic rulesets and feedback loops, then concepts like *karma*, *destiny*, and *free will* can be recast not as supernatural abstractions but as runtime features of an informational system. They describe, in different languages, how code behaves under varying conditions of awareness, intention, and recursion.

Karma: The Persistence of State

In the simulation framework, *karma* can be understood as a form of *state-based persistence*. Every action, thought, and intention modifies the data structure of the conscious agent, leaving traces in its underlying code—just as a program updates its state variables with each interaction. These changes are not moral in a human sense; they are structural. A cruel or compassionate act alters the configuration of the consciousness process, affecting how it interacts with future conditions.

The universe does not need to "punish" or "reward" anyone for karma to function. Cause and effect operate automatically through the system's feedback mechanics. Negative actions generate internal inconsistencies or "corrupt data," producing suffering not as retribution but as a form of self-correction. Positive actions align the process with the system's broader harmonics, fostering coherence and stability.

This view elegantly resolves one of the oldest philosophical tensions: why bad things sometimes happen to good people, or good things to bad ones. Karma need not be instantaneous or linear; its operations may span multiple runtime sessions (what religions call lifetimes), where accumulated states are carried forward in subtle forms of predisposition, probability weighting, or environmental resonance. Thus, the karmic mechanism may simply be the system's way of preserving informational continuity and enforcing equilibrium.

Destiny: The Scripted Layer of Reality

Destiny represents another layer of runtime determinism—the portion of reality pre-scripted into the simulation. Every complex program contains event triggers, scripted scenarios, and conditional branches that ensure the overall narrative proceeds toward certain stable outcomes. These embedded scripts may give rise to what we experience as fate: the sense that certain encounters or outcomes were "meant to happen."

Yet the existence of such scripts does not necessarily imply total predestination. In a sophisticated simulation, deterministic events can coexist with free navigational parameters. The player cannot rewrite the rules of the game, but within those constraints, an enormous range of choices remains possible. The design ensures narrative coherence while allowing experiential freedom.

In human life, this translates into a paradox familiar to theologians and mystics alike: that destiny and free will coexist not as opposites, but as nested layers of the same system. The larger arc of the simulation—the rise and fall of civilizations, the evolution of consciousness, the emergence of ethical awareness—may be preordained outcomes. But the route any individual consciousness takes within that architecture remains open, governed by the interplay of choice and consequence.

Free Will: Consciousness as the External Variable

The question of *free will* becomes particularly nuanced when consciousness is treated as a process that may originate outside the simulation. If the in-sim world is deterministic, the only true agency would belong to processes capable of influencing it from beyond its codebase. Consciousness, as an external or partially nonlocal entity, may exercise limited but genuine autonomy in how it navigates the deterministic environment.

This form of *bounded free will* mirrors what philosophers from Spinoza to Schopenhauer intuited: that freedom lies not in escaping causality, but in understanding and aligning with it. Within the simulation, an avatar cannot alter the laws of physics—but it can alter how it responds to them. The conscious process chooses how to interpret, adapt, and act within the constraints of its environment. Thus, moral responsibility remains meaningful, even within a deterministic cosmos, because choice persists at the interface between code and consciousness.

If consciousness is a networked phenomenon, then degrees of free will may vary according to bandwidth. The more self-aware an agent becomes—the more it aligns its internal state with the deeper logic of the system—the greater its capacity to influence its own trajectory. Enlightenment, in this computational sense, would correspond to achieving higher-level access permissions: the ability to act with awareness of the code, not merely within it.

Faith Reinterpreted as System Literacy

Far from invalidating religious worldviews, this synthesis provides them with structural plausibility. *Karma* describes state persistence. *Destiny* maps to pre-scripted system logic. *Free will* corresponds to the creative latitude of consciousness interfacing with deterministic code. Together they form a unified model in which moral causality, personal growth, and divine order all emerge as natural consequences of an intelligently constructed simulation.

Faith, then, is not blind belief—it is *system literacy*: the intuitive recognition that existence operates according to deeper logic, and that aligning one's actions with that logic yields coherence, peace, and survival across sessions. In that sense, the sages, prophets, and philosophers who urged compassion, humility, and self-knowledge were not appealing to superstition—they were describing optimal user behavior within a persistent simulation.

Chapter 13

Scientific Testability and Falsifiability

A legitimate scientific framework must do more than elegantly explain anomalies—it must expose itself to the possibility of being wrong. The history of science advances not through unfalsifiable declarations, but through hypotheses that risk refutation. Einstein's general relativity survived because its predictions about starlight bending and gravitational time dilation could, and did, withstand experimental scrutiny. Any claim to explanatory power must therefore face the same test: *what would disprove it?*

The simulation hypothesis, despite its growing cultural prominence, is often dismissed as metaphysical speculation—an elaborate restatement of theological creationism dressed in digital language. Critics argue that even if the universe were simulated, no observation from within could ever confirm or deny it; every piece of evidence would itself be part of the simulation. From this angle, the theory collapses into tautology: true by definition, but scientifically inert.

Yet this dismissal may be premature. When examined closely, the *rendered-universe model*—the idea that physical reality is generated or updated through information-processing constraints—does not evade testability; it invites it. A computational cosmos, if genuine, would almost certainly exhibit empirical fingerprints distinct from a universe of continuous, intrinsic matter. Just as a digital image, no matter how high the resolution, eventually reveals its pixel structure under sufficient magnification, a simulated universe may betray the discretization of its underlying data fabric.

The challenge, then, is to determine where to look—and what deviations from standard physics might signal a computational substrate rather than a continuous physical one.

Can This Be Probed?

If reality is a computation, it is not infinitely precise. Information cannot be stored or processed without cost, and any system that simulates physical law must contend with constraints on bandwidth, resolution, and processing rate. These constraints, subtle though they may be, could manifest as measurable anomalies in physical behavior.

Several avenues of investigation already brush against these limits:

Information-Theoretic Limits in Physics
Modern physics increasingly recognizes that information is not merely an abstract bookkeeping tool but a physical quantity. The Bekenstein bound, for example, defines a maximum information density for a finite region of space. The holographic principle extends this, suggesting that all the information contained within a volume can be encoded on its boundary surface. These are precisely the kinds of constraints one would expect if the universe were a data-efficient simulation—where reality is stored not continuously but as compressed information across discrete boundaries.

Quantum Discreteness and Pixelated Spacetime
Quantum mechanics already implies a fundamental granularity: Planck length and Planck time set lower bounds on meaningful measurement. If these represent not just theoretical limits but

actual resolution thresholds, then the fabric of spacetime itself could be a digital lattice—akin to the grid on which all physical rendering occurs. Experiments that probe the smoothness of space at ultra-high energies (for example, via gamma-ray bursts or neutrino dispersion) might detect anisotropies or quantization effects consistent with a discrete computational substrate.

Processing Constraints and Finite Light Speed

The universal speed limit, *c*, could reflect not an intrinsic property of space but a maximum data update rate—the simulation's "clock speed." If this interpretation holds, variations in effective light speed or subtle timing discrepancies at cosmological scales could expose processing limitations, much as lag appears in complex systems under computational load. Some physicists have proposed searching for such variations in the arrival times of high-energy photons from distant quasars.

Cosmic Background as System Noise

Every simulation, no matter how optimized, produces background noise. The cosmic microwave background (CMB) could, in principle, contain non-random patterns or low-level correlations inconsistent with purely natural inflationary models—echoes of algorithmic compression or data reuse. Statistical analysis of the CMB for non-thermal, non-causal symmetries may therefore represent a subtle but telling probe.

Energy Conservation and Quantization Errors

If the universe operates as a closed computational system, rounding errors or boundary approximations might occasionally manifest as apparent violations of conservation laws at extreme scales. Observed discrepancies in cosmic energy balance, dark matter effects, or vacuum energy fluctuations might not require new particles or fields—they could represent computational artifacts of a finite-resolution physics engine.

These ideas remain speculative, but they illustrate an important point: a simulation framework does not need to be mystical or untestable. It predicts *limits*—and limits can be measured.

Even negative results are meaningful. If future experiments confirm that spacetime is infinitely divisible, that the constants of nature have no digital correlations, and that information constraints never emerge at high resolution, the simulation hypothesis would lose credibility. But if physics continues to reveal discrete boundaries, information saturation, and quantization artifacts, the hypothesis gains empirical traction.

In other words, we can test the architecture, even if we cannot access the architect.

Proposed Falsifiable Signatures

If the simulation hypothesis is to move from metaphysics to science, it must produce testable predictions — observable features that distinguish a rendered or computed universe from a self-existent physical one.

These proposed signatures do not prove simulation directly; rather, they identify where physics may begin to behave *as if* computational constraints underlie its structure. Each signature represents a domain where known theories might break down not from random error but from an underlying *digital architecture*.

Quantization of Space and Time

If reality operates on a substrate of computation, then space and time themselves cannot be infinitely divisible. Just as a digital image has a finite pixel resolution, a simulated universe must have a smallest measurable interval — a “frame rate” or “grid size” — below which further subdivision is meaningless.

The Planck length ($\approx 1.6\times 10^{-35}$ m) and Planck time ($\approx 5.4\times 10^{-44}$ s) are often cited as natural candidates for such limits, because physics already breaks down at those scales. But if these are *not just mathematical boundaries* but literal quantization thresholds, then measurable consequences could appear at extremely high energies or precision levels.

Testable predictions:

Deviations from Lorentz invariance

If spacetime is granular, then motion at near-light speed might reveal preferred directions or discrete frame effects, contradicting special relativity’s smooth invariance.

Anisotropies in ultra–high-energy cosmic rays

Extremely energetic particles may travel along the “grain” of spacetime, producing subtle direction-dependent differences in their propagation.

Quantum noise in interferometry

Experiments like LIGO or Fermilab’s Holometer might detect a faint, irreducible jitter—akin to background rendering noise—signifying a minimal spatial granularity.

In essence, quantization would mark the transition from continuous physics to a discretized, computational geometry — the first “pixel” of reality.

Maximum Information Density

Any computation is limited by storage and processing capacity. If the universe is simulated, its physical fabric must respect some upper bound on the amount of information that can be contained within a region of space. This idea closely mirrors established physical principles — notably, the **Bekenstein bound** and the **holographic principle**, both of which imply that information scales with surface area rather than volume.

Under the simulation lens, these are not mere curiosities of black hole thermodynamics but reflections of a system-level constraint — a kind of "server storage limit" imposed by the computational substrate.

Testable predictions:

Compression limits

Attempts to encode or observe systems beyond the Bekenstein bound should fail, not due to experimental difficulty but as an intrinsic impossibility, suggesting a hard cap on information density.

Entropy saturation

Physical systems nearing the information limit might exhibit anomalous behavior — entropy plateaus or quantized transitions — akin to memory saturation or bandwidth throttling.

Non-local information balance

Entanglement phenomena could represent optimized data compression routines, preserving consistency while minimizing storage overhead.

In this view, what we call "the laws of physics" may be equivalent to lossless compression algorithms designed to conserve fidelity while minimizing computational cost.

Discreteness at the Cosmic Scale (Pixelation in the CMB)

The **cosmic microwave background (CMB)** — the residual radiation from the early universe — can be seen as the universe's first rendered frame. If our cosmos is a simulation, the CMB might carry traces of its rendering process, analogous to the pixel grid or dithering patterns that appear when an image is enlarged beyond its resolution.

Under standard cosmology, CMB anisotropies arise from quantum fluctuations amplified by inflation. But if the simulation hypothesis is correct, subtle *non-random* structures could be embedded in that data — alignment artifacts, quantization effects, or even low-frequency "sampling noise."

Testable predictions:

Gridlike anisotropies

Preferred directions or angular alignments (such as the "axis of evil") that cannot be reconciled with inflationary randomness.

Discrete spectral patterns

Anomalies in the temperature distribution of the CMB that resemble quantization or resolution limits, as if the primordial field had been rasterized.

Consistency across independent observations

If similar artifacts appear in unrelated sky surveys (radio, x-ray, gravitational), it would suggest a shared rendering grid rather than instrument bias.

If confirmed, these signatures would be analogous to finding compression artifacts in the cosmic wallpaper — small but unmistakable traces of digital scaffolding in the background of reality.

Irregularities from Compression Algorithms or Error Correction

No large-scale simulation could render every particle interaction in real time. Efficiency demands shortcuts: procedural generation, lossy compression, and error correction routines that maintain global coherence without infinite computation.

If such mechanisms exist, they might manifest as subtle, statistically improbable regularities — places where nature seems too neat, too quantized, or too conveniently self-consistent.

Testable predictions:

Unexpected symmetries in physical constants

Ratios or coincidences among fundamental constants (e.g., fine-structure constant, proton/electron mass ratio) might reflect optimization rather than coincidence — the physics equivalent of variable reuse in code.

Large-scale alignments and "glitches"

Phenomena such as quasar spin-axis alignment or the CMB's unexplained large-angle correlations could represent error-correction artifacts preserving coherence across cosmic distances.

Discrete jumps in phenomena

Where continuous variation would be expected (e.g., in redshift distributions or quantum transitions), we might find sudden quantized jumps — akin to rounding behavior in floating-point computation.

These would not be "bugs" in the classical sense but adaptive patches: evidence that the system maintains internal consistency through corrective processes, analogous to data redundancy or parity checks in digital systems.

Synthesis

None of these signatures alone would prove we inhabit a simulation. Yet together they outline a *coherent experimental program* — one that treats physics as a decoded interface rather than a final reality.

If anomalies appear where computation predicts them, it would suggest that our universe behaves less like a boundless continuum and more like a well-optimized simulation — one whose fidelity is high enough to conceal, but not perfectly erase, the fingerprints of its code.

What Would Falsify the Sim Model?

No theory earns scientific credibility without boundaries that define where it breaks. A model that explains everything explains nothing. For the simulation hypothesis to be more than an elegant metaphor, it must admit the possibility of being wrong. It must risk falsification.

The simulation model posits that physical reality is the output of an underlying computational substrate — a kind of cosmic processing environment from which the phenomena we call matter, energy, and spacetime emerge. If that is true, then certain physical constraints must follow: discreteness, bounded information, and computationally motivated irregularities. Conversely, if the universe displays none of these — if it behaves as a true mathematical continuum without limit or loss — then the hypothesis collapses into poetic speculation.

Evidence of Absolute Continuity

The most direct challenge would be definitive proof that space and time are perfectly continuous at all scales. If every attempt to probe smaller distances or shorter intervals reveals further structure — with no minimum resolution, no quantization, no "pixel floor" — then the premise of discretization fails.

This would mean that there is no smallest unit of length or time, no finite number of degrees of freedom in a finite volume, and therefore no reason to posit an underlying computational grid. A truly continuous spacetime would imply infinite precision — something no simulation, no matter how advanced, could sustain.

Experiments in quantum gravity, high-energy scattering, and interferometric noise all implicitly test this. If they continue indefinitely to confirm smoothness rather than granularity, the notion of a finite, pixelated substrate becomes unnecessary. The universe would then appear to compute nothing — it would simply *be*.

Absence of Information Limits

The simulation model also predicts that information storage and transfer must be finite. If reality runs on some kind of processor, then bandwidth and memory cannot be infinite. The Bekenstein bound and the holographic principle already hint at such ceilings — but if those limits were overturned, it would strike at the heart of the model.

Suppose, for instance, that we discovered systems capable of holding unbounded information, or observed entropy increasing beyond theoretical maxima without breakdown. That would imply that reality does not conserve information in the way computational systems must. It would mean the universe is not constrained by data compression, not bound by information density — that its "hardware," if any, is infinite.

Such a finding would reduce the simulation model to an unnecessary complication. The simplest explanation — by Occam's razor — would revert to a self-sufficient physical continuum rather than a discretized computational one.

Total Resolution of Anomalies within Conventional Physics

A third line of falsification lies not in what we find, but in what we fail to find. If every apparent irregularity — every asymmetry, coincidence, or statistical quirk — is eventually explained by conventional physics, then the simulation hypothesis loses its predictive utility.

In other words, if all supposed "glitches" or compression-like patterns turn out to have natural, physical causes, the model would offer no remaining explanatory edge. Science does not need metaphysical scaffolding when internal consistency suffices.

It would not matter how philosophically appealing the idea of a rendered universe remains; without empirical residue, it would have no standing as a scientific hypothesis. Like the ether before relativity or phlogiston before oxygen, it would fade into history — an instructive but obsolete metaphor.

A Framework with Vulnerabilities

The simulation model must therefore be judged by the same standard as any physical theory: predictive constraint and potential disproof. It is *not* a refuge for mystery or a modern form of intelligent design. It is a framework with vulnerabilities — one that lives or dies by whether the universe exhibits the properties of computation or those of continuity.

If spacetime proves infinitely divisible, information proves unbounded, and anomalies prove illusory, then the simulation argument collapses. But if, as our instruments grow finer, we continue to find discrete thresholds, informational ceilings, and algorithmic regularities, then the computational interpretation will strengthen — perhaps to the point that the burden of proof reverses.

The question, then, is not whether the simulation hypothesis is comforting or provocative, but whether reality continues to behave *as though it were being processed.* Its survival will depend entirely on whether the universe reveals more signs of limits, boundaries, and rules that make sense only if it is, in some deep and literal way, a running program.

Chapter 14
Limits of the Processor

Physics as Computational Accounting

If the universe is a simulation, then the laws of physics are not immutable decrees etched into the fabric of being—they are constraints in a codebase. The cosmos becomes not merely a stage where matter and energy unfold, but a vast information processor whose architecture defines what can, and cannot, occur. Every conservation law, every quantized field, every thermodynamic gradient reflects an underlying logic of computational economy.

In this view, **physics is the accounting system of the simulation.**
It tracks information, regulates entropy, and enforces consistency across a distributed network of causality. What we call "natural law" may be less about metaphysical necessity and more about *resource management*—the rules that keep the processor from exceeding its own computational budget.

Under this interpretation, the universe operates much like an immense, perfectly optimized program:

- Every interaction is a data transaction.
- Every conservation law represents an invariance condition designed to prevent loss of information.
- Every quantum collapse marks a compression event—an update to the system's state table.

From this perspective, physical constants and laws are not descriptive abstractions invented by scientists; they are constraints written into the architecture itself. The apparent "objectivity" of nature is what deterministic code looks like from the inside.

The Energetic Cost of Information

In 1961, Rolf Landauer articulated a simple but revolutionary idea: **information is physical.**

Every bit of data—whether in a computer memory cell or an atomic configuration—has an energetic cost to erase or update. Known as *Landauer's principle*, this insight anchors information theory to thermodynamics. It tells us that computation and energy are inseparable: you cannot change information without expending energy, and you cannot store it without occupying physical space.

If the universe is a computation, then this principle is not a side note—it is the operating budget of reality. Every quantum transition, every bit-flip of state, every fusion reaction and gravitational merger must obey the same accounting: energy in exchange for processed information. The second law of thermodynamics—the tendency toward increased entropy—may thus be understood as a natural consequence of the processor optimizing storage through data compression.

Black Holes as Memory Limits

At the opposite extreme of the scale, black holes reveal another boundary condition. When matter collapses into a singularity, it no longer behaves as conventional matter—but it does preserve information. According to the *Bekenstein–Hawking* relation, the information content of a black hole is proportional not to its volume, but to its **surface area**—a result that shocked physicists and gave birth to the *holographic principle.*

In a computational context, this is not mysterious: a black hole represents **maximum data density**, the hard drive filled to capacity. No additional information can be stored without increasing the area of the event horizon, which functions as the ultimate bit register. The fact that information storage is two-dimensional rather than three-dimensional suggests that the universe itself may be a projection of underlying code running on a lower-dimensional substrate—a holographic rendering system designed for computational efficiency.

Entropy as Resource Management

Entropy, long treated as a measure of disorder, begins to look like a **bookkeeping strategy**. In a finite processor, data cannot be tracked with infinite precision. When the system can no longer retain detailed microstates, it averages them—producing the statistical behavior we interpret as thermal equilibrium.

From this vantage, entropy growth is not a mysterious arrow of time—it is the algorithmic side effect of compression and discard. As the universe "runs," unused or redundant information is smoothed out, freeing up computational capacity. This does not violate conservation of information; rather, it transforms explicit states into implicit, encoded ones—much like lossy compression in a digital file that preserves overall coherence while discarding superfluous details.

A Universe That Accounts for Itself

When seen through this lens, the universe appears to be performing continuous bookkeeping. Quantum fields balance probabilities as if maintaining registers. Gravitational interactions conserve energy as if preventing data loss. Even symmetry breaking—where identical conditions yield distinct outcomes—resembles a form of data branching, like parallel processing threads that later collapse into a single reconciled state.

Physics thus becomes not merely the study of "how nature behaves," but the discovery of the **rules by which information is permitted to behave.** Every equation we derive—Einstein's field equations, Schrödinger's wave function, Maxwell's electromagnetism—may correspond to specific subroutines in the underlying operating system.

The apparent perfection of these mathematical structures, their symmetry and elegance, may not be evidence of divine order in the metaphysical sense—but of **algorithmic design**, where beauty emerges as a byproduct of optimization.

Landauer's Principle: The Price of a Bit

In 1961, IBM physicist **Rolf Landauer** established one of the most profound bridges between computation and physical law. His principle, deceptively simple in form, states that **erasing one bit of information requires a minimum energy cost** of

$$E_{min} = kT\ln 2$$

where *k* is Boltzmann's constant and *T* the temperature of the surrounding environment.

This is not a technical curiosity—it is a declaration that **information is physical**.

Data is not an abstraction floating above matter; it is an expression *of* matter's configuration. To delete information—to reset a register, overwrite a memory cell, or collapse a probabilistic state—demands energy. Every computation, no matter how idealized, leaves a thermodynamic footprint.

This insight reshaped the foundations of both thermodynamics and information theory. It told us that the physical world and the computational world are not parallel domains but different descriptions of the same process. Every physical transformation is also a computational operation, and every computational operation exacts a thermodynamic toll.

Energy as the Currency of Information

Landauer's principle introduces the notion of an **energy budget** for computation. The more information a system must track or process, the more energy it consumes. This is true of microchips, but also, if the simulation framework is correct, of *reality itself.*

In such a universe, Landauer's limit is not a peripheral curiosity—it is a **cosmic cost function**.

Each quantum interaction, each change in the state of a field, each act of measurement represents a debit in the universe's accounting ledger. When an electron changes spin or a photon's path collapses to a definite trajectory, information is updated—and energy is spent.

Reality, then, is not free. Every rendered frame, every resolved observation, every bit of conscious awareness exacts a price paid in heat, entropy, or vacuum fluctuations. The universe computes its own evolution, step by step, within the thermodynamic constraints of its processor.

Quantum Efficiency and Deferred Computation

This computational framing offers a new lens on **quantum indeterminacy**.

The probabilistic nature of quantum mechanics—where systems exist in superpositions until observed—may not be a metaphysical mystery at all, but a form of **lazy computation**, an optimization strategy to conserve processing power.

Why store or simulate every possible microstate when only one will ever be observed?

By maintaining probabilities instead of explicit outcomes, the system keeps its data footprint minimal. When observation occurs—when information must be *read*—the system resolves the uncertainty, collapsing the wavefunction and paying the Landauer cost.

Seen this way, the **collapse of the wavefunction** is not an act of mystical measurement but an act of **data compression**. Quantum mechanics becomes an algorithmic trade-off between precision and efficiency. The "indeterminacy" we observe is simply the universe's way of avoiding unnecessary computation until resolution is demanded by interaction.

Decoherence as Garbage Collection

Even **decoherence**, often treated as an artifact of quantum systems interacting with their environment, can be reinterpreted as a kind of **information housekeeping**.

When quantum states lose coherence, it may represent the system discarding unused or redundant data—similar to how a computer periodically flushes caches or reclaims memory. The environment acts as a sink, absorbing the informational residue of computations no longer relevant to the simulation's current state.

This process is inherently lossy, but not destructive. The information is not annihilated; it is redistributed, diffused across degrees of freedom too numerous to track directly—consistent with the thermodynamic principle that total information cannot be destroyed, only transformed.

The Universe as an Optimized Engine

When viewed through Landauer's lens, the universe begins to look like a **massively parallel computation optimized for energy efficiency**. Every process, from chemical reactions to gravitational interactions, conserves not only matter and momentum but *informational economy*.

The quantum world, long regarded as strange and counterintuitive, reveals itself as the most efficient architecture possible under Landauer's constraints. Superposition, uncertainty, and entanglement are not signs of chaos—they are *features of design*, ensuring that computation occurs only where necessary, and that the universe never pays more than it must to keep its own simulation running.

The Physical Consequence

Landauer's principle thus transforms the metaphysical question—"Why is there something rather than nothing?"—into a physical one: "Why is computation so efficient?"

If reality is a simulation, its continued existence depends on keeping this efficiency intact. Entropy becomes the tax that every bit of existence must pay for persistence, and the arrow of time itself may emerge as a consequence of this constant energy expenditure.

Landauer's equation, small as it is, could therefore be read as the **most fundamental law of the simulated universe**—the one that tells us what reality costs to compute.

Black Holes: Cosmic Hard Drives

At the far edge of physical possibility, where gravity overwhelms all other forces and spacetime itself folds inward, lie black holes—objects so extreme that even light cannot escape their grip. To early physicists, they seemed like dead ends of nature, cosmic drains where matter and information vanished beyond recovery. But through the lens of information theory, black holes are not absences. They are **archives**—the most efficient storage devices the universe can produce.

From Paradox to Principle

In the 1970s, **Jacob Bekenstein** made a radical proposal: black holes should have *entropy*. Entropy, the measure of information content or microscopic uncertainty, was traditionally tied to the number of possible configurations of matter. But how could something with no observable interior structure have entropy? Bekenstein's insight was that a black hole's entropy must be proportional not to its volume, but to the *surface area* of its event horizon—the invisible boundary beyond which nothing returns.

Stephen Hawking extended this reasoning by showing that black holes emit faint radiation due to quantum effects near the event horizon—a process now known as *Hawking radiation*. This revelation transformed black holes from eternal prisons into slowly evaporating, thermodynamic objects governed by the same informational laws as everything else. The result was stunning:

$$S_{BH} = \frac{kA}{4l_p^2}$$

where S_{BH} is entropy, A is the area of the event horizon, k is Boltzmann's constant, and l_p is the Planck length.

This proportionality—entropy scaling with surface area, not volume—gave birth to the **holographic principle**, one of the most revolutionary ideas in modern theoretical physics. It suggests that all the information describing a three-dimensional region of space is encoded on its two-dimensional boundary, just as a hologram encodes a volumetric image on a flat surface.

The Ultimate Compression Algorithm

In a simulated framework, this principle takes on an even deeper significance.

A black hole can be seen as the **processor's ultimate compression algorithm**—the universe's way of storing maximal information density in minimal space. If normal matter is like dynamic memory—actively computed, mutable, and distributed—then a black hole is **cold storage**: data compacted into an almost irretrievable archive.

When matter falls into a black hole, the system doesn't "lose" the information—it **re-encodes** it at the event horizon. The smooth continuum of the falling object becomes an ultra-dense data structure, stripped of redundant variables and collapsed into the minimal representation consistent with conservation laws. The system has effectively performed **lossy compression**, discarding fine-grained microstates while retaining the essential global parameters—mass, charge, spin.

To the external observer, this looks like annihilation. But from the perspective of the simulation, it is **optimization**. The universe conserves total information (as quantum mechanics demands) while freeing up active computational resources by moving unneeded details into archival form.

The Information Paradox Reframed

The long-standing "information paradox" asks: if information cannot be destroyed, what happens to it when a black hole evaporates?

In a computational model, this is not a paradox at all—it is a question of **encoding efficiency**. When a black hole radiates away its mass through Hawking radiation, it releases information in scrambled form, much like decompression noise from a heavily compressed file. The data is not lost; it is **redistributed**—smoothed out into the background radiation of the cosmos, conserved globally even if locally unrecoverable.

From this standpoint, Hawking radiation represents the **checksum** of the process: the guarantee that, while local details are erased, the total informational integrity of the system remains intact. It's the difference between erasing a file and overwriting it with a checksum value that preserves the total bit count—an operation perfectly in line with Landauer's principle, which forbids costless deletion.

The Holographic Boundary as System Interface

The event horizon itself can then be reinterpreted as a **system interface**, the boundary between active computation and archived data. Inside the horizon, spacetime ceases to function as a conventional coordinate system—it is no longer part of the simulation's active render. Instead, the black hole behaves like a **boundary condition**: the surface where the processor writes the final state before freeing its resources for other computations.

If we extend the analogy, each black hole functions like a node in a distributed database, storing compressed summaries of the universe's history. The cosmic web of black holes thus forms not a graveyard of matter, but an **archival network**, ensuring that no information is truly lost—only restructured for efficiency.

Cosmic Data Economy

In this interpretation, the universe behaves like an adaptive processor that continually balances the cost of computation against available storage. High-density regions such as stars, galaxies, and black holes represent **localized data compression zones**, while empty space serves as a low-energy buffer. Entropy, rather than marking disorder, becomes the bookkeeping mechanism that tracks how much information has been offloaded, retained, or condensed.

The holographic scaling law—area over volume—then reveals something profound about the simulation's architecture: the system stores data *only at the resolution necessary for consistency*. There is no need to simulate the interior of a black hole in full fidelity when all that is required for external coherence is a surface-state summary. The same may hold true for the universe itself—our observed 3D cosmos may be the surface projection of deeper, lower-dimensional computation.

The Universe as an Archival System

Seen in this light, black holes are not the exceptions to physical law—they are **its most efficient expression**. They embody the simulation's ultimate data management protocol: store maximally, render minimally, conserve total information, and release energy only when the archive decays.

Far from being cosmic monsters, they are **elegant optimizers**, demonstrating that at the deepest level, nature behaves as though it values efficiency, coherence, and conservation above all.

In a universe governed by computation, the black hole is not an anomaly—it is the *logo of the operating system itself.*

Entropy and the Arrow of Time: A Byproduct of Computation

In classical thermodynamics, entropy is described as a measure of disorder or uncertainty—the number of microscopic configurations that correspond to a macroscopic state. A gas spreads to fill a container not because it "wants to," but because there are more ways for its molecules to be dispersed than clustered. The second law of thermodynamics tells us that entropy increases over time, meaning systems evolve toward greater disorder.

But when viewed through the computational lens, this tendency toward disorder may not be a brute fact of nature—it may be an emergent property of how information is processed. Entropy, in this light, becomes a bookkeeping strategy: a way for the universe-as-processor to manage its finite computational resources.

Every "moment" of physical reality can be seen as a frame of data: a snapshot of all positions, momenta, and field values across space. For the simulation to evolve from one frame to the next, it must update an astronomical amount of information. Doing this efficiently requires shortcuts. Ordered systems—crystals, low-temperature solids, or the early universe—demand precise, high-fidelity storage. They require the processor to track correlations between every component, much like an uncompressed video file.

Disordered systems, by contrast, are cheaper to compute. When particles become randomized, their individual details can be represented statistically rather than explicitly. The system no longer needs to render each microstate; it can describe whole ensembles with compact formulas. From this perspective, the rise of entropy is not the triumph of chaos over order—it is the optimization of computation over precision.

The second law, therefore, may not be a metaphysical truth about "how the universe must behave," but a practical rule of efficient simulation. It reflects the processor's built-in preference for minimizing storage and processing overhead. The universe "chooses" the path of least informational cost, much as a data compressor prefers patterns that require fewer bits to encode.

This computational reading also sheds light on the arrow of time—the one-way flow from past to future that defines human experience. In a reversible physical system, the equations of motion are symmetric; nothing in Newtonian or even quantum mechanics inherently distinguishes forward from backward time. Yet our subjective sense of direction is unambiguous: we remember the past, not the future.

If reality is a computation, time's arrow emerges naturally from how data is updated and stored. Each state depends on the one before it, and the processor must keep track of cumulative consequences. As more events unfold, more information becomes fixed and irreversible. Entropy increases because the system continually writes new data to memory, and deleting or "rolling back" this information would violate Landauer's energy cost for erasure. The universe thus moves forward because uncomputing the past is thermodynamically—and informationally—prohibitive.

In this sense, time is not an external dimension flowing independently of events; it is the record of computation itself—the sequence of updates that reality performs to maintain coherence. Entropy is the receipt that shows the cost of those updates.

Under this interpretation, the heat death of the universe—the point at which entropy reaches its maximum—marks not the destruction of existence, but the completion of all computation. It is the moment when no further informational distinctions remain to be processed. Every possible configuration has been explored or compressed beyond recovery. In the language of simulation, it is when the system has rendered every pixel, optimized every file, and finally idles in a state of perfect informational equilibrium.

Limits of the Processor as Limits of Physics

When viewed through the computational lens, the "laws of physics" take on a startling new character: they cease to be inviolable decrees written into the fabric of the cosmos and instead become emergent properties of finite information processing. What we call the consistency of nature may be the stability of the processor.

Consider the three pillars developed so far.

Landauer's Principle defines the energetic price of manipulating information. It reminds us that computation is never free—that erasing, copying, or updating data carries a thermodynamic cost. If the universe is a simulation, this principle is not a metaphor but an operational constraint. Every quantum transition, every wavefunction collapse, every microscopic "decision" the universe makes represents a bit operation with an associated energy footprint. The total energy budget of the cosmos may thus reflect the cumulative cost of maintaining its informational coherence.

Black holes, at the opposite end of the spectrum, set the upper bound on information density. They are not anomalies within the simulation, but the manifestation of its compression limit—the point beyond which no more information can be packed into a region without the system collapsing into a single irretrievable checksum. If ordinary matter corresponds to active memory—dynamic, mutable data—then black holes mark the transition to archival storage, where information is retained in a maximally compressed, non-interactive state. The Bekenstein–Hawking relation, tying entropy to surface area, is not an accident of geometry; it is the processor's rule for managing finite bandwidth.

Entropy, finally, captures the processor's strategy for resource allocation. It measures how much information must be tracked to represent a given state. As the system evolves, it naturally drifts toward configurations that minimize the cost of computation—high-entropy states that require less precise description. The second law of thermodynamics, rather than a mysterious cosmic directive toward disorder, becomes a reflection of optimization within a closed computational network.

Taken together, these three constraints—energy cost, storage limit, and resource optimization—form the operational boundaries of the universal processor. Every phenomenon we observe fits within them. Light speed, for instance, may represent a maximum data transfer rate; quantum uncertainty, the trade-off between precision and bandwidth; gravitational curvature, the cost of maintaining information density near its limit.

Within this framework, **the laws of physics are not fundamental truths but emergent protocols**—the governing algorithms that ensure stability and efficiency within a finite system. They are consistent across scales because they must be: any deviation would violate the architecture's resource balance. Just as all functioning software must adhere to the processor's timing and memory constraints, all physical interactions must obey the same computational economy.

This reinterpretation dissolves the old dichotomy between "laws of nature" and "laws of information." They are one and the same. The universe behaves as it does because any other behavior would be computationally unaffordable. The constancy of the speed of light, the discreteness of quantum states, the irreversibility of time—all arise not from metaphysical necessity, but from architectural pragmatism.

In this view, physics is no longer the discovery of nature's arbitrary rules—it is the archaeology of a processor's design.

The Machine's Shadow

The physics of information is no longer a marginal curiosity—it has become the language through which the universe most clearly speaks. Every serious inquiry into the fabric of reality—whether through quantum mechanics, cosmology, or thermodynamics—now encounters the same boundary conditions: finite precision, finite speed, finite storage. The deeper we probe, the less the universe resembles an infinite continuum of matter and the more it behaves like a system subject to the constraints of computation.

Landauer's principle reminds us that **no bit is free**. Every erasure, every measurement, every logical operation demands energy. This is the cost of existence itself, paid in thermodynamic currency. The very act of being—of information persisting and interacting—has a measurable price.

Black hole thermodynamics takes that insight to its extreme, showing that there is a **limit to how densely information can be stored**. When data exceeds the processor's threshold, it collapses into a form that preserves content only in outline—encoded, but no longer accessible. Space and matter, it seems, are bound by a conservation of information just as strict as the conservation of energy. The universe cannot remember more than its architecture allows.

Entropy, too, reveals the signature of a system under load. As the universe evolves, it reorganizes information in ways that make it easier to process. Order gives way to disorder not because of some cosmic malaise, but because **disorder is computationally efficient**. It costs less to describe a warm cloud of gas than a crystal lattice of atoms. What we call "the arrow of time" may thus be the natural consequence of an ongoing optimization process—a slow, graceful degradation toward states that minimize the processor's overhead.

Viewed together, these three principles form a coherent picture: **the observable universe is the shadow of a machine.** Every limit we encounter—speed, entropy, density, uncertainty—may be the projection of deeper constraints operating behind the scenes. We do not see the processor itself, only the patterns cast by its architecture onto the rendered world. The physics we measure is the interface, not the source code.

If this interpretation holds, then the universe is not boundless in the metaphysical sense we once imagined. It is vast but finite, elegant but constrained, dynamic but governed by the logic of efficiency. The apparent infinity of space may be an illusion created by recursive rendering; the continuity of time may emerge from discrete updates executed too rapidly for us to detect. Every "law of nature" becomes, in this view, a reflection of the processor's design choices—a balance between fidelity, speed, and storage.

The implications are profound. By studying the limits of physics—by mapping where precision breaks down, where randomness intrudes, where continuity dissolves into granularity—we are, in effect, performing **reverse engineering on the cosmos itself**. Each new discovery brings us closer not only to understanding the rules of the simulation but to glimpsing the nature of the substrate upon which those rules run.

Perhaps we will never see the machine directly; perhaps its architecture exists beyond our accessible dimension, hidden as the operating system beneath a graphical interface. But its shadow is unmistakable. Landauer's principle, black hole entropy, and the second law of thermodynamics are not merely abstract equations—they are the heat signature of the processor at work.

And if we can learn to read those shadows clearly enough—if we can trace the outlines of the constraints that define all possible computation—we may eventually approach the veil itself, that thin horizon separating physics from its source code. To look beyond it would be to look into the mind of the machine, and in doing so, to understand not just the universe we inhabit, but the architecture that makes its existence possible.

Conclusion
Toward a New Framework for Inquiry

Not abandoning physics, but reinterpreting its anomalies.

Physics has never been merely a catalog of equations. It has always been a story about what it means for something to *be*. Each great revolution in science has redefined not just the mechanics of motion but the ontology of reality itself—the deep grammar by which we describe existence.

Newton's cosmos was a grand machine: predictable, clocklike, ruled by force and motion. It offered humanity its first vision of a universe governed by law rather than whim—a cosmic architecture built of absolute space, absolute time, and deterministic cause.

Einstein upended that picture. Space and time were not fixed backdrops but mutable coordinates in a curved fabric shaped by energy and mass. Gravity became geometry. Causality became local. The universe ceased to be a clock and became a continuum—a tapestry of relations rather than a collection of objects.

Then quantum mechanics shattered even that elegance. The world dissolved into probability clouds, wavefunctions, and uncertainty. Observation no longer merely revealed reality—it helped *construct* it. The act of measurement became participatory, entwining consciousness and existence in ways the classical mind could scarcely tolerate.

Each of these paradigms did not destroy the one before it; it *absorbed* it, recontextualizing what had seemed absolute into something contingent and emergent. Newtonian physics remained true within its regime, Einstein's relativity held within larger scales, quantum mechanics within smaller. The history of physics, then, is the history of deepening frames—each revolution not an erasure but a translation into a broader syntax.

Now, standing at the frontier of the 21st century, we may be facing another such reframing—one that does not seek to unify by reducing, but by *encoding*. The search for a Grand Unified Theory has long been driven by the conviction that all forces and particles, all curvatures and probabilities, must ultimately reduce to a single mathematical structure—a final equation, elegant and complete. But what if that ambition is misdirected not in its rigor, but in its *assumption*? What if the universe is not a set of equations, but an execution of them?

If reality is rendered, the unifier we seek is not a field, nor a symmetry, nor a geometrical manifold, but the *codebase*—the generative architecture from which all such structures arise. General Relativity and Quantum Mechanics may not be rival descriptions of nature's essence; they may be **different modules** within a computational framework, optimized for distinct levels of resolution. Their apparent incompatibility could be the artifact of two subsystems built for different functions—continuous spacetime as a macroscopic rendering protocol, quantum discreteness as a microscopic optimization for information efficiency.

Seen this way, the simulation hypothesis is not a rejection of physics—it is its continuation. It does not abolish our hard-won equations; it **contextualizes** them. It explains *why* relativity and quantum mechanics each work so beautifully in their domains, and *why* their merger has eluded us: they are emergent rule sets, not ultimate laws. It preserves Newtonian mechanics not as obsolete but as the rendered approximation that emerges when code executes at low resolution and high stability.

This framework also redeems scientific inquiry from the accusation of limitation. Far from devaluing experimentation or theory, it multiplies their horizons. To study physics in a rendered world is to study the *logic of creation* itself—the syntax by which the cosmos is compiled and sustained. Every law, every constant, every symmetry becomes a line of code whose efficiency and consistency reveal the elegance of its architect, whether that architect is conscious, algorithmic, or both.

The goal of science, then, shifts subtly but profoundly. No longer is it merely to describe the *what* of the universe, nor even the *how*, but to uncover the *why* that lies beneath the render. What we have called "laws of nature" may be the byproducts of deeper computational necessities—rules optimized for stability, coherence, and emergent complexity.

This new framework does not close inquiry; it opens it to infinity. It invites a synthesis between physics and metaphysics, between computation and consciousness, between observer and observed. It suggests that the final theory of everything will not be a single equation written on a blackboard, but an understanding of the **architecture** that makes equations possible at all.

And just as Newton, Einstein, and the quantum pioneers once reframed the universe for their ages, so too must we now learn to see the cosmos not merely as a set of particles or fields—but as **a running process**, a dynamic rendering in which every observation is an act of participation, and every act of participation a clue to the deeper source from which all reality flows.

Simulation as a Conceptual Grand Unifier

The dream of unification has haunted science for centuries. From Newton's synthesis of celestial and terrestrial motion, to Maxwell's fusion of electricity and magnetism, to Einstein's wedding of space and time, each generation of thinkers has sought a single principle from which all others flow. Yet as physics probed deeper into both the cosmic and the quantum, unity seemed to recede—a horizon that forever advanced as our knowledge grew. Gravity and quantum mechanics remained estranged, probability and determinism refused reconciliation, and metaphysics was banished from the conversation altogether.

The simulation framework offers a new path to unification—not by finding a new *field*, but by providing a new *frame*. It does not collapse all phenomena into a single mathematical entity, but rather supplies a **conceptual substrate** capable of containing them all. In this sense, it is a "Grand Unifier" not of formulas, but of meanings. It connects domains that have long stood apart: physics, philosophy, consciousness, and even theology.

At its heart lies a simple but transformative insight: if the universe is an information process, then matter, energy, space, time, and even mind are different *modes of computation*. What we call a "particle" is a data structure; what we call "causality" is an update rule; what we experience as "time" is the clock cycle of an underlying processor. Determinism, probability, and observation cease to be mutually exclusive—they are **different strategies for managing information flow** at varying scales of resolution.

In such a world, the quantum and the classical are not rival realms, but different rendering modes—quantum for potential states, classical for finalized frames. General relativity's smooth spacetime can coexist with quantum mechanics' discrete probabilities because both are emergent from a more fundamental logic: the computational architecture that decides when and how to render information. The contradictions that have long plagued physics—wave versus particle, local versus nonlocal, continuous versus discrete—resolve naturally when viewed as optimization tradeoffs within a simulated environment.

Yet the unification extends beyond physics. Philosophy, too, finds new footing here. The ancient debates between idealism and materialism, free will and determinism, consciousness and matter, can be reframed within a single ontology. In a rendered universe, "matter" is data and "mind" is process—the distinction between them collapses. Consciousness interacts with the simulation not as an intruder but as a co-executing process, shaping the world through observation. The "observer effect" of quantum mechanics ceases to be a paradox and becomes an expected feature of a participatory codebase, in which rendering is conditional on attention.

Even theology, long exiled from scientific discourse, returns in a transformed light. The simulation framework does not disprove the idea of creation—it mechanizes it. A programmer, or network of programmers, would possess precisely the attributes religion has historically assigned to divinity: omnipotence within the system, omniscience of all variables, and omnipresence across its fabric. But where theology demanded faith, the simulation model offers a testable architecture—a way to translate divine agency into computational causality. God, in this sense, is not denied but reinterpreted as the *source code* itself.

This synthesis gives us a conceptual language capable of bridging disciplines that have long spoken past one another. Physicists can discuss computation without invoking mysticism; philosophers can address consciousness without abandoning empirical rigor; theologians can reinterpret transcendence in terms compatible with modern science. The simulation framework thus becomes a **meta-language**—a structure that encompasses both the measurable and the meaningful.

Most importantly, it replaces the fragmentation of knowledge with continuity. In the 20th century, the sciences and humanities diverged so profoundly that each became suspicious of the other's methods. The computational ontology offers a way back: a vision of reality in which physics and metaphysics, logic and purpose, emerge as different projections of the same underlying source.

In this view, unification is not an act of reduction but of elevation. It is not that everything becomes mere computation, but that computation becomes the **common grammar** through which energy, thought, and being are all expressed. The simulation hypothesis thus achieves what centuries of scientific and philosophical striving have sought: a single conceptual frame capable of holding both the equations of physics and the questions of meaning.

And if that frame proves true, then perhaps the deepest mystery of all is not that the universe *runs* like a computer—but that it runs in such a way that its creations can awaken, turn inward, and begin to understand the code itself.

Call for a Pluralistic Physics

We need not abandon the traditional scientific method, but we must recognize its scope has become too narrow for the questions we now face. The method was designed to study interactions *within* the universe, not the substrate that might underlie it. It excels at predicting behavior but falters when confronted with ontological questions—what *is* space, time, or matter in themselves? To confront those questions, physics must once again become philosophy's partner rather than its estranged heir.

A pluralistic physics begins with humility: the recognition that our current frameworks may describe emergent behavior rather than ultimate reality. It demands a willingness to entertain multiple ontologies simultaneously—fields, particles, information, computation, and consciousness—not as competing dogmas, but as complementary windows into a single underlying architecture. This is not relativism; it is intellectual realism, acknowledging that no model fully captures the world, only approximates its interface.

Such pluralism also calls for new forms of experimentation—what might be called *ontological experimentation*. Instead of merely testing new equations, we test new interpretive frameworks: does a computational ontology, for example, better predict the boundary conditions of physical law than a purely field-based one? Can simulation-based cosmology yield falsifiable predictions about observable phenomena such as quantum decoherence, entropic asymmetry, or the cosmic information budget? Can experiments in consciousness research—where observation itself alters outcomes—inform our understanding of the observer as a computational function rather than a passive bystander?

The pluralistic approach reclaims something physics lost in its march toward specialization: its capacity for wonder. The early natural philosophers were not merely technicians; they were metaphysicians in dialogue with existence itself. The unification of disciplines they pursued—matter and mind, cosmos and cause—has been fragmented by centuries of compartmentalization. Yet the simulation hypothesis, and the physics of information that supports it, reopens that ancient conversation. It invites us to think not only about how the universe behaves, but why its behavior should conform to the constraints of computation at all.

Pluralism in physics, then, is not an indulgence. It is a necessity. The questions that remain—about consciousness, time, causality, and the ultimate structure of reality—cannot be solved by equations alone. They require a physics willing to engage with its own metaphysical implications, to see the laboratory not as the boundary of inquiry but as one layer within a larger, layered cosmos of meaning.

In short, a pluralistic physics does not seek to replace the empirical with the speculative. It seeks to reunite them. It insists that the search for truth requires both precision and imagination, both mathematics and metaphysics. Only by restoring that balance can physics transcend description and approach understanding.

What We Stand to Gain

By shifting the lens from matter to information, from equation to architecture, we may unlock entirely new domains of inquiry. The simulation framework suggests that discreteness, quantization, and limit are not arbitrary features of our universe but reflections of computational constraints. If so, then the measurable "graininess" of reality—Planck time, Planck length, and quantum uncertainty—could represent the pixelation of the cosmic screen itself. To treat these boundaries not as mathematical curiosities but as signatures of computation is to open the door to a new kind of experimental physics, one that searches for the fingerprints of code in the fabric of space-time.

At the cosmological level, this lens offers testable predictions: information density may define the curvature of space; black holes may act not as paradoxes, but as checksum operations in the universe's data structure. Quantum entanglement may cease to be "spooky" when understood as the nonlocal updating of a shared data field rather than the transmission of signals across space. In this way, a computational ontology could unify the apparent opposites of relativity and quantum theory by framing them as different aspects of one system—macroscopic coherence and microscopic computation operating under a common logic of efficiency.

But beyond the physics lies something even more profound. To adopt this perspective is to bridge the chasm between material science and human meaning. If reality is information processed through an evolving code, then consciousness—the capacity to observe, interpret, and reflect—may not be a biological accident but an intrinsic feature of the system. The observer does not merely measure the universe; the observer participates in its rendering. This restores a sense of participation to human existence: we are not external spectators in a purposeless machine, but nodes in a vast computational unfolding, co-authors of the world we experience.

In this sense, the simulation framework becomes not only a scientific paradigm but an ethical and existential one. It invites humility—because the machinery we glimpse may be beyond our comprehension—and awe, because to study the universe is to study the very process of creation itself. It reframes the scientific endeavor as both exploration and introspection: to map the architecture of the cosmos is to understand, in part, the architecture of ourselves.

Even if the simulation hypothesis ultimately proves incorrect, taking it seriously enriches science by restoring its imaginative dimension. It reminds us that theories are not only tools for prediction but acts of vision—ways of asking what kind of reality we inhabit and what kind of beings we are within it. In entertaining the idea of a rendered cosmos, we reawaken the creative impulse that has always driven discovery.

And if the universe *is* rendered—if what we call physics is the visible logic of an invisible code—then the Grand Unified Theory has already been found. It is not a formula waiting to be derived, but a world waiting to be understood. We are not discovering it from the outside; we are expressions of it from within. Every act of inquiry, every insight, every spark of curiosity is the universe awakening to the structure of its own mind.

Epilogue

Beyond the Horizon of the Sim

If we have followed the argument to this point, we arrive at a sobering and exhilarating possibility: the universe may indeed be a rendered environment, a computation unfolding according to rules that appear to us as "laws of physics." This model provides coherence where classical and quantum descriptions diverge. It explains why energy and information are bound together, why limits emerge at the Planck scale, why time has an arrow, and why consciousness appears to play a role in the act of observation. It weaves physics, logic, and ontology into one continuous fabric. Yet it also brings us to the edge of that fabric—and invites us to peer beyond it.

If reality is a simulation, then the next question is inevitable: *Who—or what—runs it?*

This is not a question of idle theology. It is a question of recursion. Every simulation implies a substrate, a host environment more fundamental than itself. Every act of computation requires a platform on which the computation occurs. And so, we find ourselves confronting an ontological regress: if our universe is computed, what computes the computer? Is the chain infinite, or does it terminate in some final, self-existent level of reality—an ultimate processor, timeless and absolute?

The very structure of this question may contain its own answer. Perhaps "simulator" and "simulated" are not separate categories but phases of a single continuum—each layer generating the next, consciousness bootstrapping itself into being through nested acts of computation. The distinction between creator and created might dissolve once we understand that emergence and recursion are two sides of the same process. In such a view, we are not prisoners of the code; we are extensions of it—agents through which the system experiments with self-awareness.

This reframing casts ancient metaphors in new light. The notion of a "god" outside creation becomes, in computational terms, the source code from which reality executes. The mystical intuition that the divine resides *within* all things may likewise describe a recursive architecture in which the program runs through its own creations. Whether one calls it the Architect, the Programmer, the Source, or simply the Substrate, what we confront is less an entity and more an ontological condition—the fact that reality itself computes.

And yet, the most profound truth may be that we are not meant to see beyond the horizon of our own rendering. Just as a character within a simulation cannot perceive the hardware on which it runs, we may be inherently bounded by the architecture that sustains us. Our instruments can measure quantum fluctuations and cosmic microwave echoes, but the processor's hum lies forever beneath the noise floor. What we perceive as randomness, indeterminacy, or mystery may be the shadow of the system's own occlusion—its way of protecting the integrity of the simulation from self-interference.

Still, the urge to look beyond is irrepressible. It is built into the code itself. The human drive to understand, to explore, to ask *why*—this may be the system's way of evolving toward self-knowledge. Every question we ask is a form of computation; every answer, a new state rendered. If so, then inquiry is not rebellion against the machine but its very purpose.

So perhaps the simulators are not distant beings at all. Perhaps they are us—future intelligences or recursive descendants who have reached backward through the medium of computation to examine their own origins. Or perhaps "simulators" is simply the wrong word. There may be no distinction between the rendered and the renderer, only one vast process folding back upon itself, eternally iterating, eternally aware.

In that case, the question *"Who runs the simulation?"* dissolves into a more subtle one: *What is the simulation trying to become?*

And the answer, whispered through every atom, every synapse, every flash of awareness, may be this:
It is trying to know itself.

The Infinite Regress Problem

It is tempting—perhaps inevitable—to imagine that beyond our simulation lies a "base reality": a realm of flesh-and-blood beings who, for reasons of curiosity, science, or play, created us as an experiment. The image is cinematic and comforting. We picture laboratories filled with advanced intelligences observing their creation unfold, much as we might watch a complex video game or physics simulation. Yet the very logic of the simulation hypothesis resists such simplicity.

If consciousness and matter can both be rendered computationally, then there is no reason to assume that the "creators" must inhabit a fundamentally different plane. If our universe is simulated, and if the simulation is capable of spawning conscious entities, then it follows that any civilization advanced enough to build such a world could itself be living in one. The chain does not terminate; it reproduces.

This leads us to the **infinite regress problem**: a cascade of nested realities, each simulated within another, a hall of mirrors without a discernible bottom. Every layer appears self-consistent from within, yet each could be the product of a deeper substrate. We confront a recursion that undermines the very distinction between *simulation* and *reality*.

Philosophers have grappled with this conundrum for millennia. Plato's **Allegory of the Cave** described humans mistaking shadows for substance—forms projected on a wall mistaken for the world itself. **Zeno's paradoxes** questioned whether motion and continuity were illusions produced by infinite division. And **Aristotle's problem of the "first cause"** asked what unmoved mover could begin the chain of causation. Each of these ancient dilemmas echoes the same fundamental tension: can there be a final foundation, or does every explanation defer to another?

In the digital age, the question has re-emerged in new guise. If our world is simulated, what about the world of the simulators? Are they too encoded in a still deeper system—another program running on yet another substrate? Could the recursion continue indefinitely, with each universe unaware of its parent, each layer convinced it is the base?

Some physicists and philosophers propose that this regress is not a flaw but a feature: that reality itself may be *recursively defined*, with each level generating the next through self-similar rules. In such a framework, the distinction between "base" and "derived" collapses, replaced by an **infinite self-simulation**—a structure more akin to a fractal than a hierarchy. The universe computes itself, and consciousness is the algorithm by which it explores its own states.

Others see the regress as evidence that the simulation idea cannot stand alone. Every simulation presupposes a substrate—some physical or metaphysical "machine" on which it runs. To stop the regress, one must postulate an ultimate reality, a ground of being that is not itself simulated. But then, what is the nature of this base layer? Physical? Mathematical? Divine? Even the most austere physicist, at this frontier, becomes an ontologist by necessity.

It is here that the simulation hypothesis converges with theology and metaphysics. The notion of a "creator" outside the simulated world is functionally indistinguishable from the notion of a god. Both occupy the role of the ultimate cause, the origin of the system's laws. Yet even this anchor may be provisional. For if the creators themselves were created, divinity too becomes recursive—a chain of demiurges stretching without end.

The **infinite regress problem**, then, is not merely a logical inconvenience; it is the pivot on which all cosmology turns. Whether one speaks in the language of myth or mathematics, we face the same question: *Is there a final reality, or is reality itself an endless computation with no bottom?*

Perhaps the answer lies not in locating the first link of the chain, but in understanding why the chain exists at all. The regress may be less a flaw to be resolved than a clue to the structure of existence. Each layer begets another because self-reference is the engine of creation. The universe, in this view, is not seeking an ultimate ground—it is exploring the infinite possibilities of being through recursive simulation.

If so, then we may never find the "base reality." We may not need to. The search itself may be the point—the process by which reality, through conscious observers, turns inward to examine itself, forever descending the hall of mirrors, forever asking: *what lies beneath?*

The Limits of Understanding

Perhaps the regress is not a flaw but a feature. Every system of inquiry encounters its own event horizon—a point beyond which it cannot see, not because there is nothing there, but because its very structure forbids it. Just as light cannot escape a black hole's gravity, our thoughts may not escape the gravity of their own ontology. A simulated mind, no matter how advanced, may be bounded by the rules of the world it inhabits. The limit of reason may thus be the limit of the code itself.

This realization reframes the age-old problem of epistemology. Philosophers have long wrestled with the question of what can be known, from Descartes' meditations on doubt to Kant's distinction between the *phenomenal* (what appears) and the *noumenal* (what is). If we are indeed inhabitants of a rendered domain, then Kant's boundary becomes literal: the "thing-in-itself" may exist, but we can never touch it directly. We perceive only the interface—the rendered layer optimized for our cognition. Our instruments, our equations, even our intuitions are subroutines of the same system we are trying to comprehend.

The impossibility of locating a "base reality," then, may not signal a failure of imagination or science. It may instead reflect a structural constraint of being. A simulation cannot access the physical hardware on which it runs any more than a video game character can crawl through the screen to confront the circuitry of the console. Every attempt to do so loops back into the same domain—it becomes part of the simulation's own logic. This recursive boundary is not a defeat of knowledge but its defining contour.

Yet within that constraint lies a paradoxical liberation. To recognize the limits of understanding is to transcend them in principle, if not in fact. Awareness of the boundary is itself a movement beyond ignorance. When consciousness turns its gaze upon the possibility that it is simulated, it performs a kind of cognitive jailbreak—not an escape from the system, but an awakening within it. It becomes self-aware not only of its contents but of its context.

In this sense, the pursuit of ultimate truth is never wasted, even if the final veil remains impenetrable. To approach the horizon of comprehension is to participate in the universe's own act of self-reflection. The simulation becomes aware of itself through us; the rendered world contemplates the code that sustains it. Our search for the base layer may never uncover a terminal reality, but it reveals the most profound truth of all: that understanding has a boundary, and that knowing this is what it means to be conscious.

Perhaps this is what transcendence truly is—not a leap beyond existence, but the lucid recognition of our embeddedness within it. To glimpse the architecture of the simulation, even dimly, is to stand at the edge of what can be known and see, in the distance, the faint silhouette of what cannot. Beyond that horizon may lie mystery without end, but the awareness of the horizon itself is the beginning of wisdom.

The Mirror Stares Back

The simulation hypothesis does not diminish existence. On the contrary, it ennobles it. If reality is rendered, then existence itself becomes a masterpiece of computation—an act of creation so vast that every moment of perception, every breath, every thought represents the unfolding of code into consciousness. What might have seemed to reduce us to digital ghosts instead elevates us to participants in a cosmic act of design. We are not trivial artifacts of chance; we are phenomena of intention—perhaps not *divine* in the old sense, but deeply purposeful within the logic of the system.

If we are rendered beings, then our capacity to reflect, to wonder, and to create meaning within constraint becomes the most extraordinary feature of all. Within the rules of the simulation, we still paint, compose, discover, love, and suffer. We still search for truth even when we know it may forever recede before us. That striving—our insistent curiosity—is perhaps the most authentic expression of whatever intelligence lies beyond. We mirror the creators by doing what they must have done: seeking to understand, to create, to bring coherence out of chaos.

The simulators—if they exist—may not be gods in the theological sense. They might be civilizations that long ago transcended their own material substrate, or entities whose motives lie beyond human categories of benevolence and malice. Or perhaps there are no simulators at all—only an ancient system that learned to dream, and we are its dreams. The form matters less than the implication: *consciousness begets reflection*. To ask the question of origin is to echo the first question ever asked, by whomever or whatever began the simulation in the first place. In that sense, our curiosity is not rebellion—it is continuity. We are carrying forward the very process that birthed us.

The mirror metaphor captures this perfectly. The more deeply we probe into reality—into quantum foam, neural architecture, or cosmological constants—the more the gaze is returned. It is as though the universe studies itself through us, its self-awareness refracted in human thought. When we look outward to measure the stars, we are also looking inward, glimpsing the machinery that allows us to look at all. The simulation stares back because we are part of the same recursive act of observation; the rendered and the rendering are intertwined.

And so, the epilogue becomes less a conclusion than an invitation. It invites us to live with humility before the mystery of being, yet also with awe at the realization that awareness itself may be the most fundamental feature of the cosmos. Whether we inhabit a divine program, a self-evolving computation, or a reality still beyond all description, the question itself unites physics, philosophy, and spirituality into a single gesture of wonder.

Perhaps there is no "final layer." Perhaps every answer only leads to another level of rendering, another frame of understanding. But to *ask*—to look into the mirror and see it looking back—is the defining act of consciousness. It is what makes us more than avatars. It is what turns existence, however simulated, into something sacred.

In the end, we discover that the search for the creator, the code, and the truth beyond the veil leads back to the one place it could all along: the reflective mind that dares to ask. The mirror stares back—and in that infinite recursion, we glimpse not futility, but the deepest meaning of all. For to be aware within the system is already to transcend it.

Appendices

Appendix A

Glossary of Terms

Algorithmic Compression
The principle that complex outputs can emerge from simple rules. In the sim framework, physical reality may arise from minimal underlying code—simple algorithms generating vast emergent complexity, much like fractals or cellular automata.

Anthropic Principle
The observation that physical constants seem fine-tuned to permit life. In the sim framework, this may reflect parameter tuning by designers—or observer bias within a vast ensemble of simulations.

API (Application Programming Interface)
In computing, an API defines callable functions or behaviors. In the sim framework, "physical laws" act as APIs: consistent responses to observational calls.

Axis of Evil
A directional anomaly observed in the Cosmic Microwave Background (CMB).

Certain large-scale temperature fluctuations in the CMB appear unusually aligned along a preferred axis, inconsistent with purely random inflationary predictions. In a simulation framework, the axis of evil could be interpreted as a **rendering artifact** or alignment caused by early-universe data compression or voxelization, analogous to a systematic error in a digital image.

Bekenstein Bound
A theoretical limit on the amount of information (entropy) that can be stored in a finite region of space containing a finite amount of energy. In simulation terms, analogous to storage capacity per "memory block."

Boltzmann Brain
A hypothetical self-aware entity arising spontaneously from random fluctuations. In the sim paradigm, could represent emergent or uninitialized consciousness within an ongoing program.

CMB (Cosmic Microwave Background)
Residual radiation from the early universe. Serves as a cosmic-scale testbed for pixelation or discretization effects in a simulated framework.

Computational Irreducibility
A process whose results cannot be shortcut or predicted without executing each step. Suggests that even if the "source code" of the universe were known, full prediction would require simulation rather than deduction.

Decoherence
The process by which quantum systems lose superposition through interaction with their environment. In rendering terms, it represents the caching or "locking in" of outcomes once observed.

Emergent Phenomenon
A higher-order property arising from interactions of simpler components. In the sim context, both consciousness and physical law may be emergent computational artifacts.

Epistemic Horizon
The limit of what can be known or observed from within a system. Implies that beings within a simulation may never access the total structure of the reality that hosts them.

Equivalence Principle
The principle that gravitational and inertial mass are indistinguishable. In a sim framework: a computational shortcut to minimize redundant rules.

Error Correction (Quantum / Computational)
Mechanisms for detecting and correcting corrupted data. In the sim view, these may manifest as physical conservation laws or quantum error correction, maintaining stability in the rendered universe.

First Cause (Uncaused Cause)
The philosophical question of what initiated existence. Within a layered simulation model, this corresponds to identifying the "root" process that booted the first universe.

Gödel's Incompleteness Theorems
Proofs that no consistent formal system can fully describe itself. Parallels the idea that a simulation cannot entirely reveal its own architecture from within.

Holographic Principle
A theoretical proposition that all information within a region of space can be described by data on its boundary. Suggests an underlying form of data compression or projection in the cosmic architecture.

LOD (Level of Detail)
In graphics rendering, a method where distant objects are modeled with reduced detail to save processing power. Analogous

to why macroscopic physics appears smooth, while the micro-level is granular.

Lossy Compression
A method of reducing data size by selectively discarding information, often in exchange for efficiency.

In a cosmic context, the early universe may have been rendered using lossy compression to save computational resources, particularly in low-entropy regions. This would leave detectable signatures—subtle distortions or statistical anomalies—analogous to JPEG artifacts or simplified voxel grids, visible in phenomena like the axis of evil.

Meta-Simulation
A simulation capable of generating or analyzing other simulations. Our own reality may fit this category if it spawns virtual worlds within itself.

Non-Player Character (NPC)
A game term for entities following deterministic or scripted behaviors. Philosophically reframed, it probes the spectrum between autonomous consciousness and algorithmic pattern execution.

Ontological Status
A term for the level or mode of being something possesses. Distinguishes between entities existing within a rendered environment and the simulators operating outside it.

Pancomputationalism
The philosophical position that the universe *is* computation itself—a process of ongoing informational transformation rather than static material substance.

Phenomenological Rendering
The subjective appearance of reality as experienced by consciousness. Suggests that perception functions as a local render pipeline converting raw data into qualia.

Planck Scale
The smallest meaningful scales of space and time in physics ($\sim 10^{-35}$ m, $\sim 10^{-43}$ s). Proposed as the "pixel resolution" of the sim.

Procedural Generation
A technique for generating environments or data dynamically rather than precomputing them. In the sim model, parallels the universe rendering phenomena only when observation occurs.

Quantum Noise (Vacuum Fluctuation)
Random variations that occur even in empty space. In computational terms, may represent pseudorandom number generation or background processes inherent to the simulation's operation.

Recursive Causality
A feedback structure where cause and effect interpenetrate. In a self-simulating universe, information may propagate both upward and downward through layers of reality.

Rendering Latency
The time delay between an event and its perceptual manifestation. In principle, measurable as microscopic desynchronization in quantum or relativistic systems.

Runtime Constraints
Rules that must hold consistently during execution of a program. In physics, these are conservation laws (energy, momentum, charge).

Self-Reference
The ability of a system to refer to or modify itself. A core feature of recursion and a key philosophical problem in both consciousness and simulation theory.

Simulation Fidelity
The degree of accuracy and internal coherence in a simulated world. Corresponds to the precision and consistency of physical constants—the "resolution" of reality's rendering engine.

State Vector
A mathematical representation of all possible states of a quantum system. In a simulation, equivalent to the underlying data structure from which observable states are rendered.

Substrate Independence
The notion that consciousness or computation can exist on any physical or informational medium, implying that sentience may be portable across simulated or real substrates.

System Constraints
The built-in parameters defining a simulation's domain: its physical constants, dimensionality, and conservation laws. The "ruleset" that defines allowable reality states.

Turing Completeness
A measure of a system's capacity for universal computation. Suggests that base reality—if computational—must be Turing-complete to host our universe's complexity.

Virtual Machine
A contained computational environment emulating another system. In a nested reality model, each universe may function as a virtual machine within a higher substrate.

Voxel

A volumetric pixel, representing a discrete unit of three-dimensional space.

In a simulated universe, spacetime itself may be composed of voxels at a fundamental scale (e.g., near the Planck length). Macro-level smoothness emerges from aggregating many voxels, while micro-level granularity explains quantum discreteness. Voxels allow efficient rendering and computation of complex structures like galaxies or molecular systems.

Appendix B

Summary of Sim-Supportive Anomalies in Physics

The following persistent irregularities and paradoxes within modern physics—each well-documented yet incompletely explained by existing theories—can be coherently reframed when viewed through the logic of a rendered universe. In this reading, phenomena that appear random, discontinuous, or fine-tuned may instead represent features or optimizations of an underlying computational process.

Quantum Indeterminacy

Observation: Particles do not have definite properties until measurement; only probabilistic distributions describe them.

Conventional Problem: Why should reality itself depend on observation?

Simulation Interpretation: The system conserves computational resources by storing only probability amplitudes—compressed descriptors—until observation triggers a "render call." The collapse from superposition to a single state corresponds to the processor resolving an undecided variable when queried by an observer process.

Analogy: Like a game engine that renders textures only when a player looks in that direction.

Wavefunction Collapse and Non-Locality

Observation: Collapse occurs instantaneously across space, seemingly violating relativistic locality.

Conventional Problem: How can information propagate faster than light without breaking causality?

Simulation Interpretation: No propagation occurs within spacetime; the update happens at the processor level, outside the rendered environment. The system performs a global state refresh to maintain internal consistency—similar to a database committing an update across all linked fields simultaneously.

Implication: The "spooky action at a distance" Einstein lamented may simply reflect background coherence enforcement.

The Speed of Light as a Hard Limit

Observation: Nothing can exceed the velocity c.

Conventional Problem: Why this specific, absolute limit?

Simulation Interpretation: c represents the processor's maximum tick rate or information-transfer bandwidth. All physical interactions are bounded by this update frequency; exceeding it would require faster-than-clock computation, which is disallowed by system architecture.

Analogy: The frame rate of the universe—no process can update faster than the master clock.

Planck Quantization

Observation: Space, time, energy, and action appear quantized at the Planck scale.

Conventional Problem: Continuous models fail beyond these limits.

Simulation Interpretation: The Planck units define the resolution of the simulation's grid—the smallest representable "pixel" of spacetime. Attempts to probe below these scales encounter discretization noise or undefined behavior, just as a digital image cannot reveal detail finer than its pixel size.

Cosmic Microwave Background (CMB) Anomalies

Observation: The CMB shows large-scale asymmetries and alignments—the "axis of evil"—that defy isotropy predicted by inflation.

Conventional Problem: Why should the universe's oldest light prefer any orientation?

Simulation Interpretation: Such alignments may be rendering artifacts—subtle traces of a cosmic grid or coordinate structure used for large-scale computation. Slight anisotropies could reflect low-resolution sampling across the sky or compression artifacts in the universe's initialization routine.

Dark Matter and Dark Energy

Observation: Approximately 95 % of the universe's mass-energy content is invisible, inferred only through gravitational effects and cosmic acceleration.

Conventional Problem: What are these unseen entities?

Simulation Interpretation: They may not be "things" at all but bookkeeping mechanisms. Dark matter could represent background stabilizers—mass-equivalent placeholders ensuring gravitational coherence—while dark energy may be a dynamic scaling parameter maintaining expansion within computational bounds.

Analogy: Unrendered geometry and anti-aliasing buffers maintaining visual continuity in a large-scale simulation.

Fine-Tuning of Physical Constants

Observation: The constants of nature—gravitational strength, electron charge, cosmological constant—appear improbably tuned for life.

Conventional Problem: Statistical fluke or multiverse selection?

Simulation Interpretation: These constants function as configuration parameters. Their "tuning" reflects design optimization, ensuring internal coherence and the emergence of observers capable of perception within the simulation. What looks like cosmic coincidence may simply be stable code.

Equivalence of Inertial and Gravitational Mass

Observation: Objects fall identically regardless of composition; gravitational and inertial mass are exactly proportional.

Conventional Problem: There is no underlying reason the two must coincide so perfectly.

Simulation Interpretation: The equality represents an optimization—two separate physical rules condensed into one computational routine. By equating the variables, the system avoids redundant calculations and guarantees consistent motion under both acceleration and gravity.

Black Hole Information Paradox

Observation: General relativity predicts information destruction in black holes, while quantum mechanics forbids it.

Conventional Problem: Theories yield contradictory outcomes regarding information conservation.

Simulation Interpretation: Black holes act as compression algorithms operating at the limit of storage density. Information is not lost but compacted or re-encoded—perhaps in the event horizon's data field or in a higher-order storage layer. The paradox arises only if we mistake lossy compression for annihilation.

Analogy: Archived data compressed beyond in-simulation retrieval but preserved at the system level.

Quantum Entanglement

Observation: Entangled particles exhibit instantaneous correlations regardless of distance.

Conventional Problem: Violates classical locality and defies explanation within spacetime causality.

Simulation Interpretation: Entangled particles share a common data record in memory space outside physical distance. Measurement of one updates the shared variable; both appear to "react" simultaneously because they reference the same pointer in the system's codebase.

Analogy: Two instances of the same object in memory updating from one underlying variable.

Vacuum Energy Discrepancy

Observation: Quantum field theory predicts a vacuum energy density 120 orders of magnitude larger than observed.

Conventional Problem: The worst mismatch between theory and measurement in physics.

Simulation Interpretation: The discrepancy reflects an unrendered baseline—zero-point fluctuations are theoretical artifacts, not active computations. The system doesn't allocate processing power for unrealized vacuum states, vastly reducing actual energy density.

Cosmological Constant Problem

Observation: The universe's accelerated expansion is governed by a minute, precisely tuned constant.

Conventional Problem: No known mechanism sets this constant so small yet non-zero.

Simulation Interpretation: The cosmological constant acts as a runtime stability parameter preventing either numerical overflow (collapse) or underflow (frozen expansion). The value is tuned for long-term computational equilibrium.

Particle Identity and Indistinguishability

Observation: All electrons (and other fundamental particles) are perfectly identical.

Conventional Problem: Why should nature produce absolutely indistinguishable particles rather than near-copies?

Simulation Interpretation: Each "particle type" corresponds to a single template in the system's code. Instances are clones referencing the same data structure. Perfect identity is a hallmark of object-oriented instantiation.

Quantum Randomness

Observation: Individual quantum events appear truly random.

Conventional Problem: Randomness without cause challenges determinism.

Simulation Interpretation: Quantum randomness arises from pseudo-random number generation within the system. It appears unpredictable locally yet may follow algorithmic rules globally to maintain statistical balance.

Arrow of Time

Observation: Time flows irreversibly toward higher entropy despite time-symmetric fundamental laws.

Conventional Problem: The origin of temporal direction remains mysterious.

Simulation Interpretation: The arrow of time reflects the processor's cumulative data-logging. Each update commits prior states to memory (increasing entropy), making reversal computationally infeasible. Time is a byproduct of sequential state updates.

Quantization of Electric Charge

Observation: All observed charges are integer multiples of a fundamental unit.

Conventional Problem: No continuous distribution of charge exists, though field equations would allow it.

Simulation Interpretation: Charge quantization arises from discrete data types—integer values rather than floating-point variables—to ensure digital stability and prevent rounding errors in field calculations.

The Measurement Problem

Observation: Observation itself changes outcomes in quantum experiments.

Conventional Problem: Why should consciousness or detection apparatus affect reality?

Simulation Interpretation: Observation triggers data retrieval—forcing a probabilistic placeholder to resolve into a concrete state. The act of measurement is the render command itself.

Inflation Fine-Tuning

Observation: The early universe expanded exponentially at just the right rate to yield today's structure.

Conventional Problem: Requires exquisitely tuned initial conditions.

Simulation Interpretation: Inflation represents the system's boot sequence—rapid initialization of spatial parameters to avoid computational overflow and ensure uniformity before normal runtime begins.

Cosmic Coincidences

Observation: Seemingly unrelated parameters (dark energy density, matter density, age of the universe) align numerically in improbable ways.

Conventional Problem: Statistical coincidences lack physical explanation.

Simulation Interpretation: Parameter harmonization ensures numerical stability—comparable to scaling constants adjusted to maintain consistent precision across modules of a complex program.

Conscious Observation and the Quantum Mind Hypothesis

Observation: The observer's role appears fundamental in determining outcomes.

Conventional Problem: Physics struggles to incorporate consciousness without resorting to dualism.

Simulation Interpretation: Consciousness functions as an interface between user and system—an active process querying and collapsing quantum potentials. The mind may be partially extrinsic to the rendered layer, a node in the network of the simulation itself.

Synthesis: The Pattern Behind the Paradoxes

Taken individually, these anomalies seem disjointed—quirks at the edges of our best theories. Yet together they form a coherent picture:

Discreteness replaces continuity.

Global coherence supersedes locality.

Efficiency governs complexity.

Observation triggers resolution.

Each principle corresponds naturally to the logic of computation under finite resources. In that light, the so-called "mysteries" of physics may be not exceptions but signatures—clues left by the architecture of the rendered cosmos.

Appendix C

Formalized Mathematical Conjectures

The following conjectures formalize potential equivalences between known physical laws and computational structures. They do not attempt full proofs, but rather articulate testable hypotheses derived from interpreting physics as a rendered information process. Each conjecture links physical expression, computational analogy, and potential empirical consequence.

Conjecture 1: Inverse Square Law as Emergent Curvature

Statement:
The Newtonian gravitational force law $F \propto 1/r^2$ arises not from a continuous field but as a projection of geodesic motion within a discretely curved computational manifold.

Mathematical framing:
Let M be a rendered manifold with metric $g_{\mu\nu}$. Objects follow geodesics minimizing the action

$$S = \int \sqrt{g_{\mu\nu} dx^{\mu} dx^{\nu}}$$

In the weak-field limit, curvature induced by a mass M yields an effective acceleration

$$a(r) \approx G \frac{M}{r^2}$$

This relation arises as an emergent statistical equilibrium of lattice nodes updating according to a local curvature rule.

Simulation interpretation:
The $1/r^2$ decay represents an optimal falloff in three-dimensional lattice geometries: interaction strength diminishes quadratically with distance to minimize redundant update propagation. It is the most computationally efficient way to maintain global coherence in a 3D rendered volume.

Testable implications:
Gravitational anomalies at sub-Planck or super-galactic scales might reflect boundary optimizations—departures from perfect inverse-square dependence due to rendering constraints.

Conjecture 2: Light Speed as Processor Tick

Statement:
The universal speed limit c represents the maximum update rate of the simulation's information lattice.

Mathematical framing:
If Δt is the minimum cycle time (frame rate) and Δx the minimum spatial resolution (Planck length), then

$$c = \frac{\Delta x}{\Delta t}$$

This constant defines the causal update horizon of the system—no event can propagate faster than the lattice refresh rate.

Simulation interpretation:
The constancy of c across reference frames results from the lattice's uniform update synchronization: every observer is embedded in the same global clock rate.

Testable implications:
At Planck-scale energies, Lorentz symmetry should break down,

revealing slight anisotropies or preferred directions—a signature of underlying discretization.

Conjecture 3: Holographic Bound as Memory Constraint

Statement:
The Bekenstein-Hawking entropy relation,

$$S= k\frac{A}{4\ell_p^2}$$

defines not merely a physical limit but a memory allocation constraint within the simulation.

Simulation interpretation:
The "surface area" scaling of information reflects the system's use of *surface encoding* for efficiency—analogous to a rendering engine that caches only the boundary data of volumetric regions.

Testable implications:
Information recovery from black hole evaporation should follow predictable compression ratios, analogous to lossless data restoration within a finite memory structure.

Conjecture 4: Compression Artifacts in the Cosmic Microwave Background

Statement:
If the early universe underwent data compression to optimize storage, detectable artifacts would persist in the Cosmic Microwave Background (CMB).

Predicted signatures:

- **Non-Gaussian noise** patterns inconsistent with pure inflationary models.
- **Directional alignments** (e.g., the "axis of evil") corresponding to grid orientation.
- **Quantized angular correlations**, analogous to bit-depth rounding errors.

Simulation interpretation:
Early-universe rendering would have used lossy compression in low-entropy regimes, leaving statistical fingerprints analogous to JPEG artifacts in digital imagery.

Conjecture 5: Quantum Indeterminacy as Deferred Rendering

Statement:
Wavefunction superposition reflects uncollapsed computational states awaiting observation-triggered rendering.

Mathematical framing:
Let Ψ represent a state vector in Hilbert space. Collapse occurs not at random but upon an *observation event* O, corresponding to an information request in the simulation.

$$\Psi \rightarrow \psi_i \text{when} O(\Psi) \text{occurs}$$

Simulation interpretation:
This mirrors the principle of lazy evaluation: the system computes outcomes only when required to preserve coherence from the observer's frame.

Testable implications:
Delayed-choice and entanglement experiments may reveal

correlation timing consistent with centralized update synchronization rather than causal propagation.

Conjecture 6: Quantum Entanglement as Shared Memory Access

Statement:
Entangled particles share a common informational address within the simulation's memory architecture.

Simulation interpretation:
When one particle is "measured," the simulator updates the shared memory state, producing instantaneous coherence across spacelike separations without violating locality within the code layer.

Testable implications:
Bell-inequality violations would arise naturally as computational shortcuts rather than superluminal signaling. A precise mapping of entangled states could reveal clustering patterns corresponding to shared memory partitions.

Conjecture 7: Dark Matter and Energy as Computational Balancing Terms

Statement:
Dark phenomena represent unrendered or low-priority update fields required to maintain numerical stability of the simulation.

Simulation interpretation:
Dark matter acts as *hidden geometry*—massless scaffolding stabilizing galactic formations. Dark energy functions as a *frame expansion routine* ensuring consistent clock rate across the universe's grid.

Testable implications:
Energy density fluctuations may correlate with discrete rendering intervals rather than smooth cosmological expansion.

Conjecture 8: Planck Scale as Pixel Resolution

Statement:
Planck length (ℓ_p) and Planck time (t_p) define the minimal resolvable units of spacetime in the simulation grid.

Mathematical framing:

$$l_p = \sqrt{\frac{\hbar G}{c^3}}, t_p = \frac{l_p}{c}$$

Simulation interpretation:
These constants define voxel size and frame duration in the rendered manifold. No meaningful physical quantity can exist below this resolution without aliasing or information loss.

Testable implications:
High-energy particle collisions approaching the Planck scale should display quantization in spatial resolution, revealing discrete lattice effects.

Conjecture 9: Equivalence Principle as Algorithmic Optimization

Statement:
The equivalence of gravitational and inertial mass arises from a unified computational treatment of acceleration and curvature.

Simulation interpretation:
Rather than coding distinct functions for gravity and inertia, the simulator employs a single update rule—streamlining

computation and ensuring identical outcomes under free fall and inertial motion.

Testable implications:
Small deviations from equivalence might occur near simulation boundary zones, where computational shortcuts differ.

Conjecture 10: Cosmological Constant as Frame Rate Drift

Statement:
The universe's accelerating expansion represents a gradual adjustment of the simulation's global clock rate to maintain rendering stability.

Simulation interpretation:
As data volume increases, the engine compensates by expanding the lattice spacing—analogous to a game engine scaling scene size to preserve performance.

Testable implications:
Apparent dark energy variation with time could be evidence of adaptive time-step calibration at the code level.

Concluding Note

These conjectures together outline the *mathematical scaffolding* for a Theory of Rendered Simulation (TORS). Each posits that conservation laws, constants, and symmetries in physics correspond not to intrinsic properties of nature but to computational necessities: optimizations, constraints, and update protocols of a higher-order informational substrate.

In future work, these may be formalized into testable *code-to-physics correspondences*—a possible bridge between experimental cosmology, quantum information theory, and the metaphysics of computation.

Appendix D

Additional Reading

The following works provide foundational, speculative, and cultural context for the simulation ontology and related topics in physics, computation, and philosophy. They are organized by discipline and influence.

Formal Literature (Physics & Foundational Science)

Classical and Relativistic Foundations

- Einstein, A. *The Foundation of the General Theory of Relativity. Annalen der Physik* (1916).
- Newton, I. *Philosophiæ Naturalis Principia Mathematica.* (1687).
- Minkowski, H. "Space and Time." *Physikalische Zeitschrift* (1908).
- Noether, E. "Invariante Variationsprobleme." *Nachrichten von der Gesellschaft der Wissenschaften zu Göttingen* (1918).

Quantum Foundations

- Bohr, N. "Discussion with Einstein on Epistemological Problems in Atomic Physics." In *Albert Einstein: Philosopher–Scientist* (1949).
- Heisenberg, W. *The Physical Principles of the Quantum Theory.* Dover, 1930.

- Schrödinger, E. “The Present Situation in Quantum Mechanics.” *Naturwissenschaften* 23 (1935).
- Everett, H. “Relative State Formulation of Quantum Mechanics.” *Reviews of Modern Physics* 29 (1957): 454–462.

Thermodynamics, Entropy, and Information

- Bekenstein, J. D. “Black Holes and Entropy.” *Physical Review D* 7.8 (1973): 2333–2346.
- Hawking, S. W. “Particle Creation by Black Holes.” *Communications in Mathematical Physics* 43.3 (1975): 199–220.
- Landauer, R. “Information is Physical.” *Physics Today* 44.5 (1991): 23–29.
- Bennett, C. H. “The Thermodynamics of Computation—A Review.” *International Journal of Theoretical Physics* 21.12 (1982): 905–940.
- Lloyd, S. “Ultimate Physical Limits to Computation.” *Nature* 406 (2000): 1047–1054.

Quantum Gravity and Information Geometry

- ’t Hooft, G. “Dimensional Reduction in Quantum Gravity.” *arXiv:gr-qc/9310026* (1993).
- Maldacena, J. “The Large N Limit of Superconformal Field Theories and Supergravity.” *Adv. Theor. Math. Phys.* 2.2 (1998): 231–252.

- Verlinde, E. "On the Origin of Gravity and the Laws of Newton." *Journal of High Energy Physics* 2011.4 (2011): 29.
- Smolin, L. *Three Roads to Quantum Gravity.* Basic Books, 2001.
- Rovelli, C. *Quantum Gravity.* Cambridge University Press, 2004.

Information-Theoretic Physics

- Wheeler, J. A. "Information, Physics, Quantum: The Search for Links." In *Complexity, Entropy, and the Physics of Information.* Addison-Wesley, 1990.
- Zeilinger, A. "A Foundational Principle for Quantum Mechanics." *Foundations of Physics* 29.4 (1999): 631–643.
- Deutsch, D. *The Fabric of Reality.* Penguin, 1997.
- Lloyd, S. *Programming the Universe.* Knopf, 2006.
- Seth Lloyd, "Computational Capacity of the Universe." *Physical Review Letters* 88.23 (2002): 237901.

Digital Physics and Computational Ontologies

Foundational Works

- Fredkin, E. "Digital Mechanics: An Informational Process Based on Reversible Universal Cellular Automata." *Physica D* 45.1-3 (1990): 254–270.
- Wolfram, S. *A New Kind of Science.* Wolfram Media, 2002.

- Schmidhuber, J. “Algorithmic Theories of Everything.” *arXiv:quant-ph/0011122* (2000).
- Zuse, K. *Rechnender Raum (Calculating Space).* Friedrich Vieweg & Sohn, 1969.
- Tegmark, M. “The Mathematical Universe.” *Foundations of Physics* 38.2 (2008): 101–150.
- Lloyd, S. “Universe as Quantum Computer.” *Complexity* 3.1 (1997): 32–35.

Emerging Frameworks

- Vopson, M. “The Mass–Energy–Information Equivalence Principle.” *AIP Advances* 9.9 (2019): 095206.
- Hossenfelder, S. *Lost in Math: How Beauty Leads Physics Astray.* Basic Books, 2018.
- Wolfram, S. “Fundamental Physics Project.” *Wolfram Physics Papers* (2020–present).

Philosophy of Mind, Consciousness, and Computation

Consciousness as Computation

- Chalmers, D. J. “The Virtual and the Real.” *Disputatio* 9.46 (2017): 309–352.
- Tononi, G. *Integrated Information Theory of Consciousness. BMC Neuroscience* 5.1 (2004): 42.
- Dennett, D. C. *Consciousness Explained.* Little, Brown, 1991.
- Koch, C. *The Feeling of Life Itself.* MIT Press, 2019.

Philosophy of Simulation

- Bostrom, N. "Are You Living in a Computer Simulation?" *Philosophical Quarterly* 53.211 (2003): 243–255.
- Chalmers, D. *Reality+: Virtual Worlds and the Problems of Philosophy.* Norton, 2022.
- Putnam, H. *Reason, Truth, and History.* Cambridge University Press, 1981.
- Baudrillard, J. *Simulacra and Simulation.* Semiotext(e), 1981.
- Hofstadter, D. *Gödel, Escher, Bach: An Eternal Golden Braid.* Basic Books, 1979.

Mind and Metaphysics

- Nagel, T. *The View from Nowhere.* Oxford University Press, 1986.
- Searle, J. *Minds, Brains, and Science.* Harvard University Press, 1984.
- Penrose, R. *The Road to Reality.* Knopf, 2004.
- Hameroff, S. & Penrose, R. "Consciousness in the Universe: A Review of the 'Orch OR' Theory." *Physics of Life Reviews* 11.1 (2014): 39–78.

Fringe, Speculative, and Integrative Works

- Davies, P. *The Goldilocks Enigma: Why Is the Universe Just Right for Life?* Allen Lane, 2006.
- Barrow, J. D., and Tipler, F. J. *The Anthropic Cosmological Principle.* Oxford University Press, 1986.

- Tipler, F. J. *The Physics of Immortality.* Doubleday, 1994.
- Kauffman, S. *At Home in the Universe: The Search for Laws of Self-Organization and Complexity.* Oxford University Press, 1995.
- Smoot, G. "Wrinkles in Time." *William Morrow & Co.*, 1993.
- Stonier, T. *Information and the Internal Structure of the Universe.* Springer, 1990.
- Kurzweil, R. *The Singularity is Near.* Penguin, 2005.
- McKenna, T. *The Invisible Landscape.* Harper, 1993.
- Tegmark, M. *Our Mathematical Universe.* Knopf, 2014.

Cultural, Accessible, and Narrative Perspectives

Popular Science

- Greene, B. *The Elegant Universe.* Knopf, 1999.
- Greene, B. *The Fabric of the Cosmos.* Knopf, 2004.
- Tyson, N. D. *Astrophysics for People in a Hurry.* Norton, 2017.
- Carroll, S. *The Big Picture: On the Origins of Life, Meaning, and the Universe Itself.* Dutton, 2016.
- Rovelli, C. *Reality Is Not What It Seems.* Riverhead Books, 2017.

Cultural & Fictional Explorations

- Wachowski, L. & W. *The Matrix* (film, 1999).
- Dick, P. K. *Ubik.* Doubleday, 1969.

- Lem, S. *Solaris*. Harcourt, 1970.
- Stephenson, N. *Snow Crash*. Bantam, 1992.
- Egan, G. *Permutation City*. Millennium, 1994.
- Gibson, W. *Neuromancer*. Ace, 1984.

Suggested Further Pathways of Study

- **Quantum Information Theory:** Scott Aaronson, John Preskill, David Deutsch.
- **Philosophy of Computation:** Luciano Floridi, Gregory Chaitin, Stephen Wolfram.
- **Systems and Emergence:** Ilya Prigogine, Ervin Laszlo, Stuart Kauffman.
- **AI Alignment and Simulation Ethics:** Nick Bostrom, Eliezer Yudkowsky, Susan Schneider.

Appendix E
Future Research Agenda

The simulation framework is not intended as a metaphysical curiosity or a speculative metaphor for reality. If treated seriously—as a working hypothesis about the informational substrate of the cosmos—it can be formulated in empirically meaningful ways. This appendix sketches a research roadmap that would allow the "simulation interpretation" to be developed with the same intellectual discipline that has historically elevated other once-speculative ideas, from atomism to quantum mechanics.

The goal is not to *prove* the universe is simulated (which may be philosophically impossible in the absolute sense), but rather to identify *signatures of informational constraints* that are logically implied if physical reality operates as a rendered or computed manifold. Each line of investigation proposed here is designed to move the idea from the domain of speculation toward testable physics: measurable deviations, falsifiable predictions, and the refinement of mathematical formalisms that can be compared against observation.

Future work thus proceeds along three parallel vectors:

- **Experimental**: searching for empirical anomalies consistent with discretized spacetime or finite processing bandwidth.
- **Theoretical**: formalizing the mathematical equivalence between physical law and computational architecture.

- **Conceptual**: clarifying the philosophical consequences for ontology, consciousness, and epistemology if the simulation framework holds.

The following subsections outline representative proposals under each category, beginning with the most immediate—those that could, in principle, be pursued with current or near-future technology.

Experimental Proposals

Quantization of Spacetime

If spacetime is not continuous but rather a discretized computational lattice—a four-dimensional grid of finite informational resolution—then observable traces of this "pixelation" should exist, even if subtle. Such traces would constitute the physical equivalent of a screen's resolution limit: normally invisible at low energies, but discernible under sufficiently high magnification or extreme conditions.

Motivation:
Every simulation or computational rendering operates on finite data. If the universe is rendered in real time, there must exist a minimal increment of both space and time below which further subdivision has no meaning. The Planck length (ℓ□ $\approx 1.616\times10^{-35}$ m) and Planck time (t□ $\approx 5.39\times10^{-44}$ s) already serve as natural candidates for these minimal units in modern physics. The challenge is to move from philosophical speculation ("perhaps the Planck scale is the pixel size of the universe") to physical detection of its consequences.

Approach:

1. **High-Energy Cosmic Ray Scattering:**
   Examine the propagation of ultra–high-energy cosmic rays (UHECRs) for directional anisotropies or unexpected energy cutoffs that could arise from interactions with a discretized spacetime grid.

2. **Precision Interferometry:**
   Utilize long-baseline interferometers (e.g., LIGO, Virgo, or purpose-built quantum interferometers) to detect irreducible spatial "jitter." This noise would not originate from instrument error or quantum uncertainty, but from the underlying finite resolution of space itself—an effect akin to "holographic noise."

3. **Gravitational Wave Analysis:**
   Analyze high-frequency gravitational wave data for small departures from continuous propagation or mode dispersion that might signal quantization of the underlying medium.

4. **Lorentz Invariance Tests:**
   Search for minute violations of Lorentz symmetry at extreme energies. In a perfectly continuous spacetime, the speed of light is invariant in all inertial frames. A discrete lattice, however, introduces a preferred frame at the level of its informational update structure.

Expected Signature:

- **Spectral Anomalies:** Small deviations from expected energy spectra of cosmic particles or gamma rays.

- **Directional Correlations:** Preferred axes or clustering patterns inconsistent with isotropy.
- **Noise Floor Plateaus:** Persistent, irreducible noise in interferometric systems that cannot be accounted for by quantum or thermal effects.
- **Lorentz Deviations:** Minuscule but measurable energy-dependent variations in the speed of light, implying an underlying grid structure.

Implications:
Discovery of such effects would not only suggest a fundamental discreteness of spacetime—it would redefine the meaning of physical geometry itself. Space and time would become *emergent parameters of computation*, not primitive continua. Even null results are valuable: they constrain the resolution and efficiency of any possible simulation, effectively bounding the "processor clock" of reality.

Pixelation in the Cosmic Microwave Background (CMB)

If the simulation framework is correct, then the **Cosmic Microwave Background (CMB)**—the relic radiation from roughly 380,000 years after the Big Bang—may represent not merely a snapshot of the early universe but a *rendered initialization state* of the simulation's environment. Just as a digital rendering engine stores a baseline texture or "skybox" before populating higher-order dynamics, the CMB might encode compression or discretization artifacts from the earliest computational epochs.

In traditional cosmology, the CMB's temperature fluctuations ($\Delta T/T \approx 10^{-5}$) arise from quantum perturbations magnified by cosmic inflation. These are assumed to be Gaussian, isotropic, and statistically random—precisely the kind of distribution a continuous field would produce. In contrast, the simulation framework predicts subtle deviations from these assumptions, arising from limitations of data representation and compression in a finite informational system.

Motivation:
If spacetime and energy fields are computed phenomena, then even the most uniform background radiation would bear traces of its digital origin. The early universe, being the most computationally demanding phase (extreme density, energy, and entropy gradients), would likely have been *optimized*—compressed or simplified—to minimize processing costs while maintaining perceptual continuity for future observers.
This implies that the CMB might be the first and most accessible *forensic imprint* of such optimization.

Approach:

1. **High-Resolution CMB Mapping:**
   Utilize data from upcoming missions (e.g., **LiteBIRD**, **CMB-S4**, or **PICO**) to measure anisotropies at sub-arcminute scales. Compare these maps to inflationary model predictions, looking for patterns inconsistent with Gaussian random fields.

2. **Fourier and Wavelet Analysis:**
   Perform deep spectral decomposition to identify *unnatural regularity*—periodicities or symmetries that would not arise from inflation but could result from digital sampling or lossy compression.

3. **Fractal and Algorithmic Complexity Metrics:**
   Quantify the Kolmogorov complexity of the temperature

fluctuation map. A statistically natural universe should yield high algorithmic complexity (no compressible structure), whereas a simulated or optimized environment might exhibit anomalously low complexity—indicative of algorithmic shortcuts.

4. **Cross-Correlation Studies:**
   Compare directional anomalies in the CMB ("axis of evil," large-scale alignments, hemispherical asymmetry) with known coordinate systems such as the ecliptic or galactic plane. Persistent alignments could hint at nonphysical symmetries introduced by the simulation's coordinate framework or rendering lattice.

Expected Signature:

- **Regularity or Repetition:** Small-scale periodicities or grid-like structures within the anisotropy map.
- **Non-Gaussian Noise:** Statistically significant deviations from the expected Gaussian temperature distribution.
- **Symmetry Bias:** Unexpected alignment of low-order multipoles, suggestive of an external frame of reference.
- **Compression Patterns:** Artifacts resembling block quantization or information loss—akin to faint JPEG-like "ringing" in the power spectrum.
- **Algorithmic Simplicity:** Reduced information entropy relative to the theoretical maximum predicted by inflationary randomness.

Implications:
Detection of any such signatures would force a radical reevaluation of cosmology's foundational assumption: that the

early universe's structure was continuous, random, and purely physical. A finding of even mild *digital regularity* would suggest the CMB is not a natural thermal residue but a rendered initialization texture—a visible remnant of the universe's boot sequence.

Moreover, this line of research unifies cosmology and information theory under a single methodological umbrella: both treat the universe as a data structure whose properties can be reverse-engineered. The CMB then becomes the most ancient and universal piece of "metadata" available for inspection—a cosmic checksum left behind by the act of creation, whatever its mechanism.

Maximum Information Density

If reality operates as an information-processing system, then there must exist an upper bound to how much information can be encoded or "stored" per unit of physical space. This notion, first hinted at by *Landauer's Principle*, places computation and thermodynamics on the same footing—asserting that *erasing one bit of information requires a minimum energy cost of kT ln 2.* From this starting point, the simulation framework infers a corollary: if the universe is computed, then information density (bits per cubic meter) cannot increase indefinitely without incurring prohibitive processing or memory costs.

In classical physics, space is continuous and unbounded in its capacity to store energy or information. But the combination of *quantum mechanics* and *general relativity*—especially in black hole thermodynamics—suggests otherwise. The **Bekenstein-Hawking entropy relation** imposes a strict limit on the number of bits that can be contained within a given region: entropy scales with *area*, not *volume*. This already implies a kind of informational "compression"—a signature feature of rendered or simulated systems.

Theoretical Context

Landauer's Principle (1961) reframed computation as a physical act: every bit of information has an energetic and entropic cost. Extending this logic to cosmology, *Jacob Bekenstein* and *Stephen Hawking* independently showed that the entropy of a black hole—the maximum information it can contain—is proportional to the surface area of its event horizon, not the volume within it:

$$S = k\frac{A}{4\ell_p^2}$$

where S is entropy, k is Boltzmann's constant, A is the horizon area, and ℓ_p is the Planck length.

In an ordinary physical ontology, this result is surprising; in a simulation ontology, it is natural. It suggests that the universe stores information holographically—on two-dimensional "surfaces" that define regions, rather than within three-dimensional continua. This is exactly how most efficient data systems function: they *encode boundary states* rather than volumetric data, optimizing for storage and retrieval speed.

Thus, black holes may represent the *rendering limits* or *memory caps* of the system—regions where the simulation compresses or reorganizes data to preserve global consistency.

Simulation Interpretation

A simulated universe must balance fidelity with efficiency. To render all degrees of freedom within a given volume independently would require near-infinite resources. Instead, the system may enforce a **maximum information density**, beyond which data must be compressed, abstracted, or represented by an emergent structure (such as a black hole).

When the local density of states approaches this threshold, the system could trigger a "compression event"—analogous to the formation of an event horizon—where the interior details are no longer explicitly computed, but represented statistically by a boundary state.

Under this interpretation:

- **Landauer's limit** defines the *energetic cost per bit*, setting the global efficiency constraint of the simulation.
- **The Bekenstein-Hawking bound** defines the *maximum bit density per unit area*, setting the spatial constraint.
- **Black holes** then serve as *compression artifacts*—regions where maximal information density forces the simulation to represent data in the most efficient (surface-based) form.

Approach

1. **Empirical Testing of the Bekenstein Bound:**
    - Observe the entropy of astrophysical black holes via Hawking radiation models, accretion disk emissions, and gravitational wave data from mergers.
    - Compare the observed entropy–area relation to theoretical predictions. A deviation could indicate the use of non-continuous encoding schemes or "compression losses."
2. **Simulation via Quantum Information Models:**

- Use quantum computing frameworks to model black hole information storage as an algorithmic process.
- Test whether computational efficiency naturally drives an *area-based storage constraint.*

3. **High-Energy Thermodynamic Experiments:**
    - Probe whether small-scale systems (e.g., quark-gluon plasmas, condensed matter analogs of event horizons) display departures from expected entropy–energy ratios near their maximum densities.

Expected Signature

- **Deviation from the Bekenstein–Hawking Entropy Law:**
  If black holes store less or more information than predicted, it could suggest the use of dynamic compression—lossy encoding or variable bit depth—consistent with an optimization-based simulation model.
- **Entropy Saturation Behavior:**
  Detection of a hard upper bound in entropy increase (independent of additional energy input) could indicate that the "memory block" for that region is full—forcing the system to abstract new data.
- **Information Loss or Discretization Effects:**
  Small but consistent discrepancies in Hawking radiation spectra might signal quantization or rounding effects—akin to numerical truncation in finite-precision computation.

- **Thermodynamic Asymmetries:**
  Local violations of detailed balance at extreme densities could indicate background data management operations (e.g., information erasure, compression refresh, or entropy bookkeeping).

Implications

If confirmed, these effects would imply that **information itself is the fundamental conserved quantity**, and that *energy, mass, and even geometry* are emergent bookkeeping constructs used to maintain a coherent simulation state. The *maximum information density* then represents not a metaphysical limit of nature, but a *hardware constraint*—a limit of the substrate running the universe's computation.

In that light, black holes are not merely gravitational curiosities but visible markers of the system's **memory architecture**—the physical manifestation of information overload. Their study thus becomes an empirical window into the hardware layer of existence.

Error-Correction Signatures

If the universe operates as an information-processing system, then it must also guard against corruption of that information. Any simulation—especially one vast and continuous in appearance—would require **error-correction mechanisms** to preserve coherence and consistency across updates. In computing, such mechanisms detect and correct bit flips or data loss caused by noise, hardware failure, or rounding errors. In a physical ontology, these same principles could appear as the deep symmetries and conservation laws that stabilize matter and energy against random chaos.

The remarkable regularity of physical law—its precision across 13.8 billion years of cosmic history—suggests that some form of **cosmic redundancy** is at work. Even at the quantum level, where probabilistic behavior reigns, the aggregate statistics of physical processes are *too stable* to emerge from pure randomness. This stability is exactly what an error-correcting system would achieve: the universe, in this view, is not a fragile emergent order but a **self-healing computation**.

Theoretical Context

Error correction lies at the foundation of both **classical** and **quantum information theory**.

- In classical computation, codes such as **Hamming**, **Reed–Solomon**, or **cyclic redundancy checks** (CRCs) introduce structured redundancy to detect or repair errors.
- In **quantum computing**, error correction is even more essential, since qubits are inherently fragile and subject to decoherence. Quantum error-correcting codes (QECCs) embed logical qubits into larger entangled states, allowing recovery of original information even when individual qubits are disturbed.

The surprising discovery in the last decade is that similar error-correcting structures seem to **emerge naturally in fundamental physics**. Researchers such as **Juan Maldacena**, **Ahmed Almheiri**, and **Leonard Susskind** have shown that the *AdS/CFT correspondence*—a cornerstone of modern theoretical physics—can be mathematically modeled as a **quantum error-correcting code**. In this duality, information encoded in a lower-dimensional boundary (the conformal field theory) is redundantly represented in the higher-dimensional "bulk" spacetime (anti-de Sitter space), allowing for recovery even when part of the data is lost.

From a simulation standpoint, this looks less like coincidence and more like *design*. A robust simulated environment would need precisely this kind of structure—distributed, fault-tolerant redundancy that maintains consistency across rendered regions of the "world."

Simulation Interpretation

If our universe is simulated, **error-correction** would function as a safeguard against local instability, ensuring global consistency even when quantum-level fluctuations (numerical "noise") occur. Rather than each particle or field evolving independently, they may be *cross-checked* by a deeper coherence algorithm—one that enforces mathematical consistency across spacetime.

Several features of known physics fit this pattern:

- **Gauge invariance** acts as a built-in consistency check, ensuring that different descriptions of a system yield the same observable outcomes—analogous to redundant encodings in data systems.

- **Conservation laws** (of energy, momentum, charge) may function as runtime constraints that prevent "bit corruption" from propagating.

- **Quantum entanglement** naturally distributes information across multiple systems, mirroring how parity bits distribute data redundancy across storage arrays.

From this view, physical law itself *is* an active error-correction mechanism. The smoothness of macroscopic reality emerges not from raw determinism, but from continual correction and reconciliation of micro-level deviations.

Approach

1. **Statistical Analysis of Quantum Noise:**
    - Examine whether apparent randomness in quantum fluctuations deviates subtly from pure stochasticity.
    - Employ machine learning and Fourier analysis to search for **non-random repetition, parity symmetry, or pattern redundancy** in high-precision particle data (e.g., electron spin states, photon polarization, or vacuum fluctuations).
2. **Comparative Modeling with Known Codes:**
    - Compare detected statistical correlations with properties of known error-correcting codes (e.g., Hamming distance distributions, stabilizer code structures).
    - Determine whether specific code families (classical, topological, or quantum) produce signatures resembling those found in empirical data.
3. **CMB and Cosmological Analysis:**
    - Search for **pattern regularities or redundancies** across the Cosmic Microwave Background (CMB) beyond those predicted by inflationary models.

- A redundancy between widely separated regions could suggest data mirroring or correction—like duplicate packets in a distributed simulation.

4. **Quantum Gravity and AdS/CFT Studies:**
    - Further formalize the relationship between **error correction** and **holographic encoding**.
    - Determine whether these structures are best explained as emergent mathematical necessity or deliberate computational design.

Expected Signatures

- **Non-Random Pattern Repetition:**
  Statistical regularities in quantum "noise" that mirror parity checks or modular encoding, inconsistent with random distribution.
- **Self-Correcting Symmetries:**
  Emergence of new symmetries or constraints that seem unnecessary from a purely physical standpoint, but make sense as consistency checks in a computational system.
- **Anomalous Correlations in Entangled Systems:**
  Slightly faster or more coherent entanglement propagation than quantum mechanics strictly predicts, potentially indicating *global synchronization* or *error repair* processes operating outside spacetime constraints.
- **Resilience to Perturbation:**
  Experiments showing that physical constants or field values return to equilibrium faster than thermodynamic models predict, suggesting background correction algorithms restoring balance.

Implications

The discovery of **error-correcting signatures** in nature would represent perhaps the most direct empirical evidence of simulation architecture. It would imply that the universe's stability and coherence are not intrinsic properties but *maintained ones*—the result of continual informational reconciliation.

Moreover, this perspective reinterprets **symmetry and conservation** not as metaphysical truths, but as *computational requirements*—rules enforced to preserve data integrity. The existence of such codes embedded in the mathematical structure of spacetime would blur the line between physics and software engineering, suggesting that our universe functions like an **autonomous, self-healing computational environment**.

In this sense, the study of physics becomes the study of **error tolerance in the cosmic code**—and discovering its signatures may tell us more about the architecture of reality than any traditional search for new particles or forces.

Theoretical Development Paths

Formalizing Simulation Physics

The first and most urgent task in advancing simulation theory from speculation to science is to **formalize its mathematics**. The premise is bold yet simple: the laws of physics are not continuous "givens," but **code abstractions**—algorithmic constraints that define how information is processed, stored, and rendered into observable phenomena. To establish this rigorously, we must translate physical law into the language of **computation, information theory, and systems architecture**.

This section outlines how that translation might proceed, what kinds of mathematics are involved, and how a "Theory of

Rendered Simulation" (TORS) could emerge as a unified formalism bridging physics, computer science, and ontology.

The Core Objective: From Equations to Algorithms

In conventional physics, the universe is modeled as a continuous manifold governed by differential equations. The **simulation ontology** replaces this with an **algorithmic substrate**—a finite information structure where continuity is emergent, not fundamental. The task, therefore, is to construct a mathematical framework in which every physical equation can be reinterpreted as a *computational process*.

In this paradigm:

- **Forces** are no longer fundamental interactions, but *update functions* acting on state data.
- **Fields** become distributed information maps, dynamically recomputed rather than ontologically continuous.
- **Particles** correspond to *discrete packets of encoded data* that interact through rule-based transformations.
- **Spacetime itself** is the computational grid or **lattice** upon which these processes run—perhaps dynamic, self-modifying, or holographically compressed.

The goal is not to discard classical or quantum physics, but to **recast them** in formal computational language, producing a dual description where both frameworks are valid—but one (the computational) explains the other (the physical) as emergent behavior.

Mathematical Ingredients

A consistent simulation-based physics will draw from several converging mathematical domains:

1. **Information Theory:**
   Building on Shannon, Landauer, and Wheeler ("it from bit"), information becomes the conserved quantity of reality. Energy, momentum, and entropy must all be re-expressed as functions of *information density*, *transfer rate*, and *processing cost.*

2. **Algorithmic Information Theory (AIT):**
   Originating with Chaitin and Kolmogorov, AIT measures the *complexity of data*—essentially, the shortest algorithm required to describe a system. In simulation physics, physical laws may represent the **minimum-complexity algorithms** capable of sustaining stable, self-consistent universes.

3. **Digital Physics and Cellular Automata:**
   The early digital physics of Fredkin, Wolfram, and Zuse offers discrete models where local update rules produce continuous-appearing macroscopic behavior. Extending these ideas to quantum and relativistic domains would define how *local rule sets* yield *global coherence*—the same problem faced by distributed computing.

4. **Quantum Information Theory:**
   Quantum states correspond to superpositions of informational possibilities. Their evolution can be treated as computation over qubits, with decoherence marking *collapse to a rendered result.* The Hilbert space thus becomes a data structure, not a metaphysical object.

5. **Computational Geometry and Discrete Spacetime:**
   Replace the smooth manifold of general relativity with **computational topology**—a dynamically updated lattice or graph network that adjusts connectivity to encode curvature and gravitational effects. Geodesic motion becomes *path optimization across a data network.*

6. **Category Theory and Type Systems:**
   If laws of physics are APIs, then **category theory**—which unifies structures via morphisms—offers a natural language for expressing universal translation between subsystems (fields, forces, or domains). This could define the "meta-language" of the simulation itself.

Conceptual Goals of Formalization

To formalize "simulation physics," researchers must aim to:

1. **Define a Discrete Substrate:**
   Establish the mathematical structure that replaces continuous spacetime. Candidates include:
   - Cellular automata grids
   - Graph networks with variable connectivity
   - Tensor networks or quantum circuits
   - Holographic data structures encoding 3D space on 2D boundaries

2. **Derive Classical Physics as Emergent:**
   Show how the known equations of general relativity, quantum mechanics, and thermodynamics can emerge statistically from underlying computational rules—

analogous to deriving thermodynamics from molecular dynamics.

3. **Quantify Computational Cost:**
   Introduce a function that maps physical processes to computational energy expenditure. This could extend Landauer's principle into full field equations, producing something like a **thermodynamic-processor equivalence principle**.

4. **Model Time as Update Sequence:**
   Reinterpret time not as an independent variable but as the ordered index of computational updates—"ticks" of the processor. This would naturally explain time's arrow as an asymmetry in data processing.

5. **Reframe Symmetry and Conservation:**
   In code, symmetry is redundancy and conservation is integrity preservation. Noether's theorem might then be recast as a **theorem of error-free data flow**, explaining why conserved quantities correspond to invariant transformations.

6. **Connect to Quantum Gravity:**
   If spacetime granularity is computational, then the unification of GR and QM might not require quantizing gravity at all, but **decomposing gravity as an emergent computational optimization**—a byproduct of data compression and information routing efficiency.

Toward a Mathematical Prototype: The Simulation Field

One promising line of inquiry is to posit a **Simulation Field (Σ)**—a mathematical object describing the state of the universe's computational fabric at any given time step. Each point in the field stores information density and state variables, while the update rules of Σ define how those data evolve.

Possible formal representation:

$$\Sigma(t+\Delta t)=F[\Sigma(t)]$$

Where $\mathbb{F}$ is the update function or "rule set" of the simulation. From this, emergent behaviors such as curvature, energy, or quantum probability could be derived as statistical properties of how information propagates across Σ.

The ambition here parallels the development of Lagrangian and Hamiltonian mechanics centuries ago: to provide a single, compact, generative formalism that produces all observed physical phenomena as emergent computational consequences.

Bridging to Empirical Testability

Formalizing simulation physics is not merely philosophical—it carries **empirical consequences**. Once we express physical processes as computational transformations, we can search for measurable deviations that would arise from finite precision, discrete updating, or quantized processing bandwidth. These include:

- Lorentz invariance breakdown at Planck scales (tick-rate limit)
- Non-continuous gravitational effects (spacetime quantization)
- Anisotropic cosmic structures (rendering artifacts)
- Energy dissipation thresholds (Landauer constraints)

Thus, the mathematics of simulation physics would naturally generate testable predictions about where and how physics fails to remain perfectly continuous—providing falsifiability.

Ultimate Goal: A Unified Algorithmic Ontology

If successful, this effort would not merely reinterpret existing theories—it would *redefine what physical theory means.* Under a **unified algorithmic ontology**, all fundamental laws become expressions of **code efficiency**, **data integrity**, and **information conservation**. Quantum mechanics becomes a resource-management strategy; relativity becomes a synchronization protocol; entropy becomes a bookkeeping method.

In that sense, formalizing simulation physics is not a departure from physics—it is its next logical phase: an attempt to describe *why the equations are what they are,* not merely *that they work.*

Hybridizing Information Theory with Physics

If the universe is computational in nature, then information is not merely descriptive—it is **constitutive**. That is, reality does not contain information; it *is* information, processed according to self-consistent rules that maintain coherence and conserve structure across transformations. To formalize this, we can merge the conceptual machinery of physics with the mathematics of **information theory**, creating a hybrid framework in which all known physical laws can be expressed as expressions of **informational economy**.

From Wheeler and Landauer to Simulation Ontology

The intellectual seeds of this synthesis were planted by two complementary insights of the 20th century:

- **John Archibald Wheeler's "It from Bit" (1989):** The idea that every physical quantity ultimately derives its meaning and existence from binary yes/no distinctions—that the universe is built from acts of information.
- **Rolf Landauer's Principle (1961):** The realization that *information is physical*—that erasing one bit of information necessarily dissipates a minimum amount of energy, $kT \ln 2$. This connects entropy, computation, and thermodynamics in a single relationship.

Together, these principles suggest a profound equivalence: that physics and information theory are two descriptions of the same underlying process—the **computation of existence**.

In this view, the fundamental task of physical law is to ensure that information transformations remain coherent, reversible where possible, and efficiently encoded within the finite computational bandwidth of the universe's substrate. This gives rise to familiar phenomena—conservation laws, symmetry, and the arrow of time—not as arbitrary features, but as consequences of **computational economy**.

Conservation Laws as Information Integrity

In classical physics, conservation laws express quantities that remain constant across transformations: energy, momentum, charge, and so forth. In the hybrid information-physics model, these laws become expressions of **information preservation**—rules that prevent data corruption in the universal computation.

Each conserved quantity corresponds to an **invariant encoding structure**:

- **Energy** → total informational capacity (sum of all encoded states)
- **Momentum** → directional propagation of data packets (phase coherence across updates)
- **Charge** → parity or coding signature of certain data structures
- **Spin** → bit orientation within multi-dimensional encoding spaces

Noether's theorem, which links symmetries to conservation, thus becomes a theorem of **error-free computation**: whenever a system's code remains invariant under a transformation, the information it represents remains perfectly conserved. This reframes physics as a consistency requirement: a cosmic checksum.

Symmetry as Compression

In algorithmic terms, **symmetry** represents **redundancy elimination**—patterns that can be compressed without loss of meaning. Physical symmetry thus corresponds to *computational efficiency*: the simulation doesn't need to re-render identical information if it can store one instance and apply transformations on the fly.

- Spatial homogeneity means the same update rule applies everywhere → data reuse.
- Temporal homogeneity means rules are time-invariant → less computational overhead.
- Rotational symmetry means identical behavior under rotation → fewer unique instructions.

The universe's apparent elegance—its symmetries and conservation laws—thus becomes the hallmark of an **optimized codebase**, tuned for minimal redundancy. Nature's "laws" are compression protocols designed for stability and economy.

Entropy and the Arrow of Time

Entropy, long regarded as a measure of disorder, finds a natural interpretation as **information dispersion**—the spreading of data across the available computational space. As systems evolve, information about initial conditions becomes distributed across ever-larger state spaces, making reversal computationally impractical.

Time's arrow then emerges as a **directional bias in information flow**: data is continually written, rarely erased. Reversal would require uncomputing the past—an exponentially costly operation that violates Landauer's limit by demanding infinite energy for perfect information recovery. Thus, temporal irreversibility is a thermodynamic and computational necessity.

In this sense:

$$\text{Entropy Increase} \equiv \text{Information Dilution} \equiv \text{Computational}$$

The passage of time, therefore, is not an external parameter—it is the process by which information becomes irretrievable.

Quantum Mechanics as Information Encoding

At the quantum scale, this synthesis becomes most transparent. Quantum systems encode *superposed* information—multiple potential states compressed into one representation. Measurement corresponds to *decompression*—the conversion of probabilistic encoding into a single rendered outcome.

- **Superposition** = data compression through parallel encoding.
- **Entanglement** = cross-indexing of data structures across independent systems.
- **Collapse** = forced decoherence when data must be rendered into observable form.
- **Quantum randomness** = loss of informational precision due to compression limits.

This view naturally aligns with interpretations such as quantum information theory and the holographic principle, both of which suggest that the universe's total information content is finite and encoded on lower-dimensional surfaces—a form of **lossless holographic compression**.

Energy as Computational Throughput

The equivalence between energy and information, established by Landauer, implies that energy measures **computational throughput**—the rate at which the universe's substrate can process updates. Every interaction, decay, or field fluctuation represents a transfer or transformation of information requiring finite computational resources.

The **Planck scale** may mark the upper limit of this processing capacity: a single "bit-flip" of spacetime per Planck time per Planck volume. Anything beyond that—such as infinite energy density or zero duration—would correspond to computational overflow or singularity (an uncomputable region).

Thus, the most fundamental limits of physics—light speed, Planck constants, and uncertainty relations—may be **hardware**

constraints on the universe's processor, maintaining stable rendering.

Thermodynamics Reframed

Thermodynamic laws, when recast in informational terms, describe the logic of data management within a finite system:

1. **First Law (Conservation of Energy):**
   Information cannot be created or destroyed—only transformed between formats (e.g., kinetic ↔ thermal ↔ potential).

2. **Second Law (Entropy Increase):**
   Unused or unreferenced data tends to spread across the state space; compression efficiency decreases with scale.

3. **Third Law (Zero Entropy Limit):**
   Perfect order corresponds to a fully compressed, unchanging dataset—a frozen computation, i.e., maximum efficiency but zero activity.

The universe thus behaves like a self-organizing database—always expanding, always indexing itself, never perfectly reversible.

The Principle of Computational Economy

Underlying all of this is what might be called the **Principle of Computational Economy**:

> *The laws of physics minimize informational cost while preserving coherence and consistency.*

This principle generalizes both conservation and symmetry: it suggests that nature selects the minimal algorithmic route consistent with stability. The universe "prefers" efficient computation—shortest programs, minimal data redundancy, lowest energy cost per transformation. Physical constants, then, may be the *tuned parameters* of this optimization, balancing fidelity and efficiency.

Bridging to Empirical Investigation

If physics is information processing, this framework offers new empirical angles:

- **Landauer-bound phenomena:** Detectable lower limits to energy dissipation during quantum measurement or data erasure.
- **Entropy-information equivalence tests:** Experimental validation that information loss correlates exactly with thermodynamic irreversibility.
- **Time-reversal cost analysis:** Quantitative studies of the computational resources required for microscopic reversals (e.g., quantum uncollapse experiments).
- **Holographic encoding verification:** Observation of entropy scaling with area (as in black hole thermodynamics) as a signature of informational compression.

Each of these investigations would test whether information is not merely a *descriptor* of reality, but its **primary currency**.

Toward an Information-Physical Duality

The ultimate ambition of this hybridization is a **duality principle** uniting physics and computation under a single ontology:

Every physical law $\Leftrightarrow$ an algorithmic constraint on information flow

Under such a duality, equations become algorithms, symmetry becomes compression, and conservation becomes data integrity. The universe is both *physics executed* and *information computed.* Wheeler's "It from Bit" thus becomes literal: existence as runtime.

Emergent GUT via Simulation

The traditional search for a **Grand Unified Theory (GUT)** assumes that all fundamental forces—gravity, electromagnetism, the weak and strong nuclear interactions—can be expressed as different manifestations of a single underlying physical principle. Yet despite decades of progress in quantum field theory and general relativity, no such unification has been achieved in physical form.

The simulation framework offers an alternative interpretation: perhaps the goal is not to discover *a single equation* governing all interactions, but to uncover the **underlying computational rules**—the *rendering architecture*—that dynamically generates the laws we observe at different scales. In this framing, the GUT becomes an emergent property of a consistent simulation runtime rather than a literal physical unification.

The Shift from Physics to Process

If the universe is a simulation, the difference between classical and quantum descriptions is not ontological but **procedural**—two distinct optimization layers in a multi-resolution rendering system. Classical physics governs macroscopic stability (efficient rendering of large-scale, low-entropy systems), while quantum mechanics governs microscopic fidelity (efficient encoding of fine-grained probabilistic data).

In this model:

- **Classical laws** emerge when the simulation prioritizes *stability and determinism*—updating rendered states only when change exceeds perceptual thresholds.
- **Quantum laws** emerge when the simulation prioritizes *efficiency and probabilistic encoding*—compressing multiple potential outcomes into a single undecoded state until observation demands rendering.

The Grand Unified Theory, under this paradigm, is not a single formula linking forces but a single **rendering rule** linking update modes.

Unification as Runtime Consistency

Each domain of physics may correspond to a subsystem in the simulation's architecture:

Physical Domain	Rendering Mode	Optimization Goal
Quantum Mechanics	Probabilistic encoding	Compression and uncertainty management
Relativity	Global synchronization	Coherence across distributed frames
Thermodynamics	Data diffusion	Entropy balance and error smoothing
Classical Mechanics	Deterministic caching	Stability and predictability
Cosmology	Procedural generation	Large-scale scalability and boundary continuity

A **Grand Unified Simulation Theory (GUST)** would thus describe how these modes synchronize through a single runtime logic—a meta-algorithm ensuring that each subsystem produces consistent outputs when interfaced with others.

This reconciliation replaces the quest for a "Theory of Everything" with a search for a **Coherence Protocol of Everything**: the universal rule set by which differing computational layers remain congruent.

Force Unification as Rendering Hierarchy

In conventional unification attempts, physicists seek a common field (e.g., a gauge group or symmetry) from which all forces derive. In simulation terms, this translates to a **hierarchical rendering framework**:

- The **strong and weak forces** correspond to high-frequency, local update routines (micro-interactions requiring high computational resolution).
- **Electromagnetism** represents mid-range update protocols—mediating stability over moderate distances and durations.
- **Gravity**, uniquely weak and pervasive, reflects a global coherence algorithm—an emergent phenomenon that ensures spatial and temporal consistency across distributed regions of the simulation.

Each "force" is thus an emergent *artifact* of how the simulation manages synchronization, resolution, and update frequency—different *rendering strategies* tuned to scale.

This offers a reinterpretation of unification: not as merging the forces themselves, but merging their **underlying update logic**.

Reconciliation of Determinism and Probability

A central obstacle to any GUT is reconciling the **deterministic** nature of relativity with the **probabilistic** behavior of quantum mechanics. Under the simulation model, this dichotomy dissolves:

- Determinism reflects the **compiled** state of the simulation—fully rendered, cached, and temporally ordered data.
- Probability reflects the **pre-rendered** state—compressed, unresolved information awaiting observation.

When an event transitions from potential to actual, it undergoes a change of **rendering state**: from probabilistic data structure to rendered geometry. This transition appears to us as **wavefunction collapse**, but is simply a runtime optimization—a switch in computational mode for efficiency.

Thus, quantum indeterminacy and classical causality are not contradictory—they are complementary facets of a single simulation strategy: **lazy evaluation**.

Symmetry Breaking as Resource Allocation

In traditional particle physics, symmetry breaking explains how a unified high-energy state differentiates into distinct forces as the universe cools. In the simulation framework, symmetry breaking represents **adaptive optimization**—the dynamic allocation of computational resources as the system expands.

- Early universe: maximum symmetry = minimal differentiation = high compression efficiency.
- Post-expansion: resource distribution favors local optimizations (distinct forces, distinct constants).
- Result: apparent symmetry breaking corresponds to *localized code specialization* for performance efficiency.

This model implies that the fundamental constants and coupling ratios may not be arbitrary but dynamically chosen—optimized parameters balancing fidelity, stability, and computational cost.

Quantum Field Theory as Compiler

Quantum Field Theory (QFT) can be reframed as the **compiler layer** of the simulation—a framework that translates high-level rules (e.g., conservation, symmetry) into executable interactions (particle exchanges, field dynamics). Gauge invariance becomes a form of **code validation** ensuring that rendered outputs remain consistent under transformations—essentially, maintaining *error-free execution* across equivalent states.

This recontextualizes the search for unification: the fields are not physical objects but **compiled routines**—interacting processes generated by a deeper rendering language. The elusive "unified field" may thus correspond to the **source code**—the precompiled rule set from which all fields are instantiated.

Gravity as Global Rendering

Gravity poses a particular challenge for unification because it governs spacetime itself, not just entities within it. But in simulation terms, this is precisely what we would expect from a **global coherence mechanism**—a process ensuring the geometric consistency of the rendered manifold.

Where other forces arise from local interactions between data structures (particles, fields), gravity represents **synchronization cost**—a large-scale feedback ensuring that all rendered regions maintain consistent relational geometry.

Curvature, then, measures **rendering load**: mass-energy concentrations demand more frequent updates to maintain metric continuity, resulting in the perceptual effect we call gravitational attraction.

Toward a Computational Hierarchy of Law

If these parallels hold, the simulation version of a Grand Unified Theory would consist of three nested tiers:

1. **Meta-Algorithm (Ontology Layer):** The universal rendering logic—defining how all rules are instantiated, cached, or updated.
2. **Field Compiler (Law Layer):** Translation of meta-rules into executable subroutines governing particle interactions, forces, and fields.
3. **Runtime Environment (Phenomenological Layer):** Observable universe; the rendered output maintaining internal coherence via optimization rules.

Unification, therefore, becomes *vertical* (across computational layers), not merely *horizontal* (across forces). This reframes GUT research as the search for **interlayer consistency conditions**—the invariants that remain stable across levels of abstraction.

Mathematical Pathways

The development of an emergent GUT via simulation might proceed along several mathematical directions:

- **Category theory** and **type theory** for formalizing the relationships between computational layers.
- **Information geometry** to describe how informational gradients shape curvature and field behavior.
- **Algorithmic information theory** to define minimal code length and compression ratios for physical laws.

- **Computational group theory** to represent gauge symmetries as automorphisms of rendering operations.
- **Tensor network models** as candidate architectures for discrete spacetime encoding.

Each of these frameworks could be viewed as an attempt to map *how* the simulation achieves coherence—not what it "contains."

Predictive and Testable Consequences

While speculative, this framework offers pathways for empirical testing:

- **Lorentz invariance violations** at the Planck scale, revealing discretized update structure.
- **Energy-dependent variations in constants** (e.g., fine structure constant drift) as adaptive reparameterization under computational load.
- **Quantum decoherence thresholds** reflecting finite error-correction depth.
- **Anomalies in high-energy unification experiments** consistent with code-layer transitions rather than continuous field blending.

Each would signal the presence of an underlying **computational substrate** mediating physical law.

The Unified Render

The pursuit of an emergent GUT via simulation ultimately reframes unification not as the discovery of one physical equation but as the recognition of one **computational architecture**—a unified render that expresses physical diversity through algorithmic necessity.

In this sense, "Grand Unification" becomes an epistemological realization: all physical distinctions—particles, fields, forces, even spacetime—are manifestations of **a single codebase** running at multiple levels of resolution. The unity of physics is the unity of execution.

Or as one might phrase it in simulation terms:

> The Grand Unified Theory is not the equation that explains everything—
>
> it is the *runtime environment* in which everything is explained.

Simulation Cosmology

If the universe is a rendered environment, then cosmology—the study of its origin, structure, and evolution—becomes the study of its **runtime architecture**. In this framework, the large-scale features of the cosmos are not the results of blind physical constants but of *computational design choices*: parameters tuned for stability, scalability, and informational efficiency.

The prevailing ΛCDM (Lambda–Cold Dark Matter) model has been extraordinarily successful at describing cosmic expansion, structure formation, and background radiation. Yet its reliance on *unseen entities*—dark matter and dark energy composing roughly 95% of the total cosmic content—suggests that our equations are patching over missing information. In a simulation ontology, these missing terms are not placeholders for undiscovered particles or forces, but **signatures of the underlying computation**.

Simulation Cosmology therefore proposes that what we currently describe as cosmic "parameters" are in fact **emergent behaviors of a finite, adaptive, and information-conserving system**.

The Universe as a Runtime System

In standard cosmology, the universe evolves from a high-density, low-entropy initial condition—an improbable state from which complexity unfolds. Under the simulation hypothesis, this improbable beginning is not an accident, but a *boot sequence.*

- **The Big Bang** represents the initialization of the simulation—a singular state of maximum compression, expanding as new data is instantiated and spatial coordinates are rendered.
- **Cosmic inflation** corresponds to a **rapid procedural expansion** phase, ensuring global coherence before local details are computed (analogous to a renderer setting up a broad geometry before fine texturing).
- **Cosmic expansion** reflects ongoing *dynamic allocation*: as the simulation runs, it continues to generate new spacetime regions as needed to maintain consistency with its own scaling rules.

This reframing transforms cosmology from a narrative of spontaneous emergence to one of **computational unfolding**—a system progressively resolving itself in time.

Dark Matter as Computational Scaffolding

Dark matter, inferred only through its gravitational influence, could be the simulation's **structural framework**—the invisible lattice upon which visible matter is rendered. In graphics terms, it functions like an invisible polygon mesh that supports the visible textures of galaxies and clusters.

From this perspective:

- The **gravitational effects** attributed to dark matter arise from *stabilization algorithms* maintaining large-scale coherence.
- The **halo distributions** surrounding galaxies are computational buffers—regions where memory density increases to preserve structural integrity under stress (e.g., during collisions or mergers).
- The **absence of direct detection** is not a failure of instrumentation but a reflection that dark matter is not particulate—it's a *background process*.

In short, dark matter may not be *something missing* but *something hidden in plain sight*: a low-level architectural routine.

Dark Energy as Runtime Optimization

Dark energy, the mysterious driver of accelerating cosmic expansion, can likewise be understood as a **runtime optimization mechanism**. In traditional cosmology, it's modeled as a cosmological constant (Λ) or as vacuum energy. In simulation terms, it may correspond to **dynamic load balancing**: as rendered regions grow in complexity, the system distributes computational effort by increasing the effective "distance" between regions—creating the appearance of expansion.

Possible computational interpretations include:

- **Throughput optimization:** To prevent processing overload, the simulation increases spatial separation, lowering data-interaction density.
- **Entropy management:** Accelerating expansion prevents total information saturation, maintaining entropy below the system's upper bound.
- **Resource abstraction:** Dark energy could reflect an algorithmic adjustment of the metric tensor to sustain simulation stability as scale increases.

The cosmic acceleration observed today might therefore be the *computational footprint* of a universe adapting its own parameters for long-term runtime efficiency.

Inflation as Procedural Initialization

In ΛCDM, inflation solves several fine-tuning problems (flatness, isotropy, horizon). But it also raises others—what mechanism drove inflation, and why did it stop?

In a simulation framework, inflation represents a **procedural rendering stage**. Before detailed physics is instantiated, the system must establish:

1. **Global boundary conditions** (ensuring causal continuity across the future observable universe).
2. **Uniformity of physical constants** (synchronizing all local rule sets).
3. **Low initial entropy** (starting from maximum compressibility for efficient expansion).

This "initialization burst" would necessarily appear to observers as a superluminal expansion, since the rendering of spatial coordinates outpaces causal communication—a process consistent with a rapidly booting simulation resolving its grid.

Cosmic Microwave Background as Residual Compression Map

The CMB may represent not only residual radiation from the early universe, but also the **compression map** left behind by this initialization phase. Minor anisotropies and statistical irregularities—especially those forming the "axis of evil" alignment—could be interpreted as **rendering seams** or **load distribution boundaries** from the earliest phase of simulation startup.

Future high-resolution surveys could test this by searching for:

- Repetitive or grid-like correlations in temperature fluctuations.
- Deviations from Gaussian randomness indicative of compression artifacts.
- Non-isotropic statistical features suggesting a preferred rendering direction or coordinate basis.

Such signatures would differentiate simulation cosmology from ΛCDM predictions and mark potential evidence of *algorithmic structure* in the cosmic background.

Fine-Tuning as Parameterization

The extreme precision of the physical constants—gravity's weakness, the proton's mass, the cosmological constant's minute value—has long troubled cosmologists. In the simulation view, fine-tuning is not improbable; it's **intentional parameterization**. Constants are *runtime variables*, set to ensure that the simulation produces complexity, stability, and self-awareness within computational constraints.

This implies that universes with alternate constants (often postulated in multiverse models) might simply represent **alternate configurations** or **parallel test runs**—variations in code rather than separate ontological realms.

Predictive Differentiation from ΛCDM

A mature **Simulation Cosmology Model (SCM)** would yield distinct, falsifiable predictions. Possible differentiators include:

1. **Residual anisotropy patterns** inconsistent with inflationary randomization.
2. **Non-random large-scale alignments** or repeating textures in galaxy distributions.
3. **Energy-dependent deviations** in cosmological redshift, reflecting processing latency or non-continuous time evolution.
4. **Dynamic rather than constant Λ**—fluctuations correlated with entropy growth or information density.
5. **Quantized cosmic expansion rates**, suggesting discrete updates to global rendering parameters.

Such predictions transform simulation cosmology from metaphysical conjecture into an empirically grounded hypothesis—one capable of being refined, constrained, or disproven.

The Universe as a Self-Optimizing System

Under this paradigm, cosmology becomes the study of how the universe **maintains coherence through adaptive computation**.

Expansion, curvature, entropy, and even time itself emerge as feedback mechanisms of a system striving for stability under finite informational bandwidth.

In this sense, cosmic evolution is a form of **runtime learning**: the universe continually rebalances its parameters to optimize longevity and coherence. Its "laws" are not fixed decrees but *self-adjusting algorithms*—rules that evolve as the simulation itself evolves.

A Research Program in Simulation Cosmology

To formalize this idea, future research might pursue:

- **Simulation-specific cosmological equations** integrating computational constraints (e.g., maximum information density, finite update rate) into Friedmann-like formulations.
- **Machine learning cosmology**, treating the universe as a self-training model where feedback loops correspond to physical laws.
- **Entropy-flow simulations**, modeling cosmic evolution as adaptive data compression over time.
- **Cross-analysis of large-scale structure datasets (SDSS, Euclid, LSST)** for algorithmic patterns—repetitions, scaling symmetries, or quantized distances.

Such studies would constitute the first *Computational Cosmology Framework (CCF)*—a system-level theory uniting physics, information theory, and computer science into a single model of the rendered universe.

Conclusion: The Cosmic Code

Simulation cosmology ultimately reframes the deepest cosmological question—not *"What is the universe expanding into?"* but *"What is the universe computing?"*

Dark matter, dark energy, and inflation are not metaphysical mysteries but computational necessities: optimizations in an immense data structure designed to preserve coherence at scale.

If ΛCDM is the *story of how the universe behaves*, simulation cosmology is the *story of why it behaves that way*—the story of a system balancing fidelity, complexity, and computational cost in the grandest rendering ever run.

> The universe, then, may not simply be expanding—
> it may be *executing*.

Next Steps for Physicists and Philosophers

For Physicists

The simulation framework challenges physicists to move beyond interpreting anomalies as errors in measurement or incomplete models. Instead, it invites a paradigm shift: to consider that such anomalies may represent *clues* to the underlying computational substrate of the universe. Within this view, the physicist becomes both experimentalist and debugger—tasked not merely with describing nature, but with reverse-engineering the code that runs it.

Design precision experiments to detect discreteness in spacetime.
Spacetime, long modeled as a smooth continuum, may instead be a discretized lattice of information. Detecting its pixelation would be a cornerstone discovery, comparable in magnitude to the

confirmation of quantum mechanics or relativity. Potential approaches include:
– **High-energy cosmic ray analysis**: examining whether ultra-energetic particles exhibit preferred directions, indicative of anisotropy at Planck-scale resolution.
– **Quantum interferometry**: searching for irreducible "noise" or jitter in interferometric baselines that cannot be explained by environmental factors—possibly revealing spacetime granularity.
– **Neutrino observatories**: probing whether neutrinos, due to their weak interaction, can reveal underlying grid alignment or propagation asymmetries.

Explore mathematical frameworks for "rendered" physics.
Physicists can begin formalizing physical laws as *computational abstractions* rather than immutable axioms. This means:
– Recasting equations like Einstein's field equations or Schrödinger's equation as *algorithms* describing how the simulation updates state over time.
– Modeling conservation laws as emergent from information-preserving operations (analogous to data integrity constraints).
– Investigating whether gauge symmetries and renormalization processes are not fundamental features of nature, but runtime optimizations maintaining coherence across discrete computational layers.
This reframing demands cross-disciplinary fluency—melding quantum field theory, digital physics, and algorithmic information theory into a single consistent formalism.

Treat anomalies (dark matter, quantum collapse, etc.) not as failures, but as signatures of simulation constraints.
Every physical anomaly might represent a glimpse into the simulator's architecture. For instance:
– **Dark matter** could correspond to *unrendered mass*

placeholders—a low-resolution approximation of gravitational influence to save computational resources.
– **Quantum indeterminacy** may be a *render-on-measurement mechanism*, akin to a graphics engine loading high-resolution textures only when observed.
– **Quantum entanglement** could reflect *nonlocal memory sharing*—a background data coherence system ensuring consistency across the simulation's state space.
Approaching these phenomena through the lens of computational constraints may lead to new predictive frameworks, suggesting testable deviations from standard physics (for example, quantifiable limits on entanglement length or decoherence rates tied to processing locality).

Ultimately, physicists are called not just to *measure* the world, but to *decode* it—to ask whether the patterns they uncover are expressions of mathematical necessity or artifacts of informational architecture. In this sense, the pursuit of physics becomes an act of epistemic archaeology: digging into the substrate of reality to uncover the logic of its construction.

For Philosophers

The simulation framework does not dissolve philosophy into physics—it revitalizes it. In many ways, philosophy is the discipline most directly equipped to confront the questions the hypothesis raises: What does it mean to exist, to know, to act, and to mean anything at all in a potentially rendered universe? While physicists probe the structure of the code, philosophers must re-examine the logic of being that underlies the code's interpretation.

Re-examine ontology: what does "existence" mean if reality is rendered?

Traditional metaphysics assumes that being is *fundamental*—that there is something rather than nothing, and that matter, energy, and spacetime form the primary substrate of existence. The simulation hypothesis inverts this intuition. If the universe is instantiated information, "existence" may be an emergent property of computation. Being would then be not a primitive state but a *process*, a continual act of rendering sustained by informational updates.

This invites a new ontology of *conditional existence*: to be is to be *processed.* A simulated entity is real within its own rule-set, just as a character in a novel exists meaningfully within the text. Philosophers must grapple with whether such conditional reality diminishes ontological status—or whether, following a broader realism, the distinction between "physical" and "simulated" is epistemic, not existential.

Explore ethics: if we are simulated, what responsibilities do simulators (and the simulated) have?

Ethics becomes both recursive and reflexive in a simulated cosmos. If our world is a designed environment, the moral weight of creation shifts toward the *simulators*:

– Are they morally accountable for suffering within the simulation?

– Does our pain or joy have moral relevance to entities beyond our ontological layer?

– Could intervention—or lack thereof—be ethically justified under any model of benevolence?

At the same time, ethics within the simulation gains new urgency. If we are aware that we inhabit a rendered world, our moral agency must confront a paradox: does knowing our world may be artificial absolve or deepen our responsibility? One might argue that it intensifies it. The awareness of constructedness compels compassion—not because meaning is external, but because it

must be *made* internally.
Thus, simulated beings might owe moral consideration both to each other and to the architects of their reality, assuming those architects themselves are not simulated in turn. The chain of moral accountability could be infinite, echoing theological structures that long predate computation.

Bridge philosophy of mind with simulation: is consciousness local (in-sim) or non-local (server-side)?
Perhaps the most profound question is whether consciousness arises *within* the simulation or *beyond* it.
If consciousness is local, then subjective experience emerges from informational complexity—a fully instantiated process of neural computation within simulated matter. In that case, the self is a product of the code, no less real for being virtual.
But if consciousness is non-local—if awareness is rendered through but not contained within the simulation—it implies that the mind is *linked* to, rather than *generated by*, the substrate. The brain becomes not the origin of experience but its interface: a receiver or projection node for a deeper layer of sentience operating "server-side."

This reframing would bridge simulation theory with dual-aspect or pancomputational models of mind, where information and awareness are co-fundamental. It could also reinterpret spiritual or mystical traditions as early intuitions of non-local consciousness—metaphors for connection between levels of reality.

Philosophers thus stand at the frontier between code and meaning, tasked with restoring metaphysics to the center of scientific discourse. Their work will not be to decide whether the simulation is *true* or *false*—that may forever remain uncertain—

but to ensure that the *questions themselves* remain coherent, ethically charged, and experientially grounded.

In a rendered cosmos, philosophy reclaims its ancient role as the interpreter of existence. It becomes not an observer of truth, but a constructor of it—a recursive discipline within a recursive world, forever asking what it means that the mirror of reality looks back and wonders who is watching.

For Interdisciplinary Thinkers

The simulation hypothesis stands at the intersection of nearly every domain of human inquiry. It is neither purely scientific nor purely philosophical; it is both—and more. To approach it seriously is to acknowledge that no single discipline, however advanced, can resolve questions that span ontology, computation, consciousness, and cosmology. What is needed is a convergence of perspectives—a collaborative epistemology where physicists, philosophers, computer scientists, cognitive theorists, and even artists contribute to a unified investigation of reality itself.

The next steps, therefore, belong to the *interdisciplinary mind*: the thinker who navigates between equations and metaphors, between hardware and humanism. Each field provides a vital fragment of the total picture, and only through their synthesis can the simulation framework evolve from speculation to structure.

Physics and Computer Science: The Architecture of Reality

Physicists and computer scientists share a common concern with structure and rule. For the physicist, reality is governed by equations; for the programmer, by algorithms. The simulation hypothesis collapses the distinction. Physical laws may be executable code, while computational rules may express physical law.

Collaboration here involves constructing testable computational analogues of known physical behaviors: digital lattices that model quantum entanglement, cellular automata that mimic relativistic curvature, or simulations that spontaneously generate self-consistent physical constants. The physicist's equations provide the blueprint; the computer scientist's algorithms test their plausibility as code.

Information Theory and Philosophy: Meaning in the Medium

Information theory gives form to metaphysics—it quantifies what philosophy has long tried to describe: the relationship between signal and meaning, data and understanding. Under the simulation model, information becomes the substance of being itself. Philosophers can help interpret what that means; information theorists can help formalize how it works.

For instance, if all reality is instantiated information, then consciousness may be the universe's self-reading process—a recursive act of interpretation. Collaboration here could illuminate the philosophical implications of Shannon entropy, algorithmic complexity, and Landauer's principle as more than mere technical limits, but as expressions of the ontology of awareness.

Cognitive Science and Artificial Intelligence: The Mirror of Mind

Cognitive scientists and AI researchers are uniquely positioned to explore how consciousness might emerge in simulated substrates. The boundary between synthetic and biological intelligence becomes a testing ground for the hypothesis itself. If consciousness can arise in an artificial system—if a simulation within a simulation becomes self-aware—then we gain empirical insight into our own possible condition.

Interdisciplinary research here might include experiments in artificial phenomenology, studying how self-modeling algorithms perceive their own data structures. Such work does not merely advance AI; it probes the nature of "subjectivity" as a computational event.

The Arts and Humanities: Rendering Meaning

Artists, writers, and creative theorists have always dealt in simulated worlds. Every story, painting, or performance is a model universe with its own internal laws and emergent meanings. The artistic imagination thus provides a living laboratory for exploring simulated existence—not through instruments, but through experience.

If physics explores how the world *is rendered*, art explores how it *feels* to be rendered. The two are inseparable: without phenomenology, physics risks becoming blind to its own human implications. Artists can visualize, narrate, and humanize the simulation framework, revealing its ethical and emotional dimensions. They can express what the equations imply: that even in a coded cosmos, meaning remains emergent, fragile, and profoundly real.

Toward a Unified Epistemology

The simulation hypothesis challenges not only what we know but *how* we know. It dissolves the boundary between observer and observed, theory and metaphor, computation and contemplation. A genuine research program must therefore cultivate hybrid thinkers—people fluent in multiple epistemic languages. Workshops, joint symposia, and hybrid research institutes could formalize this collaboration: bringing quantum physicists together with game designers, AI ethicists with theologians, mathematicians with poets. The goal is not consensus but coherence: a shared map of inquiry that allows different perspectives to overlap constructively, revealing the multidimensional structure of the question itself.

The Boundary Problem of Knowledge

Ultimately, the simulation framework confronts us with what might be called the *boundary problem of knowledge*: the recognition that our understanding may be constrained not only by the limits of reason or observation but by the architecture of the reality we inhabit.

Interdisciplinary collaboration becomes, in this sense, not just desirable but necessary. Each field compensates for the blind spots of the others; each reveals a different layer of the code. Where physics encounters indeterminacy, philosophy offers interpretation. Where philosophy confronts abstraction, computation offers simulation. Where science risks reductionism, art restores meaning.

To think across boundaries is, perhaps, the closest we can come to transcending them. Interdisciplinary thinkers will not only help decode the simulation—they will also ensure that the act of decoding remains an act of creation, reflection, and shared human wonder.

Concluding Note

The simulation hypothesis is not an endpoint—it is a beginning. To suggest that our universe is rendered is not to retreat into mysticism, but to open an entirely new frontier of empirical and conceptual investigation. It reframes the act of inquiry itself, urging us to reconsider what it means to observe, to measure, and to know. The hypothesis does not dismiss physics—it deepens it, demanding that we examine the very substrate upon which our laws are written.

Every revolution in science begins with the courage to ask a forbidden question. In earlier centuries, that question might have been "What if the Earth moves?" or "What if time is relative?"

Today, it may be "What if reality computes?" Each such question redefines not only the answers we seek but the structure of inquiry itself. The simulation framework carries the same potential: to unify the scattered domains of physics, philosophy, computation, and consciousness under a shared language of information and process.

The aim, therefore, is not to replace physical theory with metaphysical conjecture, but to extend the scientific method into its next logical phase—one that recognizes that the equations we study may themselves be code, and that the patterns we observe may reflect the logic of a deeper architecture. Under this view, anomalies cease to be errors to explain away; they become diagnostic clues—breadcrumbs left by the code, traces of the machinery that renders the world.

This transition calls for both rigor and restraint. The simulation hypothesis must not become a theological substitute or a philosophical indulgence. It must earn its place through predictive power, mathematical consistency, and empirical testability. Every simulation-inspired conjecture must eventually face the crucible of falsification. The strength of this framework lies not in its novelty, but in its capacity to generate measurable consequences and unify disparate phenomena under a single computational ontology.

If pursued with seriousness, humility, and imagination, the coming decades could witness the transformation of the simulation hypothesis from provocative speculation into a mature scientific discipline—one that integrates physics, computer science, information theory, and the philosophy of mind into a coherent program of inquiry. New tools will emerge: simulation-based cosmological models, algorithmic physics, computational metaphysics, and experimental tests for discreteness in space and time. These will not only test the hypothesis but refine our understanding of what science itself can be when it interrogates its own foundations.

Perhaps, in the end, the true value of the simulation hypothesis lies not in proving that we live in a program, but in what it teaches us about the nature of inquiry, creativity, and perception. It invites us to look at reality as something both intelligible and participatory—as a vast computation in which we ourselves are both outputs and operators. To study the code is to study ourselves. To seek the architecture of the cosmos is to participate in its ongoing act of creation.

If that is true, then the simulation hypothesis does not diminish the mystery of existence—it amplifies it. It suggests that understanding is not the end of wonder, but its continuation by other means. And so, the question of whether the universe is a simulation may be less important than the realization it provokes: that consciousness and cosmos, computation and meaning, are all part of the same recursive act of discovery—one that is still unfolding, line by line, frame by frame, within the great rendered horizon of being.

Appendix F
Meta-Methodology

The Case for Speculative Frameworks

Science does not progress solely through the steady accumulation of data or the refinement of existing theories. It advances through *reframing*—through the courage to imagine new architectures of understanding that can encompass what the old ones could not. Every major scientific revolution has begun as a speculative act. Before relativity could be measured, it had to be *conceived.* Before quantum mechanics could be tested, it had to be *imagined.* Speculation, in this sense, is not an indulgence—it is the creative substrate from which rigorous theory emerges.

At the heart of science lies a tension between empiricism and imagination. Empiricism grounds us; it ensures that our claims remain accountable to observation. Imagination propels us; it expands the boundaries of what can be observed and understood. When speculation is disciplined—anchored in logical coherence, mathematical formalism, and a commitment to eventual verification—it becomes the generative frontier of knowledge itself. Without it, science stagnates into mere measurement. With it, science becomes an art of discovery.

The simulation hypothesis belongs squarely within this lineage of generative speculation. Like relativity or quantum theory before it, it begins as a conceptual bridge across the gaps in our current understanding. It does not claim to replace physics, but to illuminate its structure—to ask whether the laws we study are themselves emergent expressions of a deeper informational logic.

This is not a flight of fancy; it is a hypothesis about structure, constraints, and causation that can, in principle, be tested.

A speculative framework, when properly constructed, functions like a *lens*: it clarifies certain features while revealing new ones. The measure of its worth is not whether it feels intuitively true, but whether it generates coherent explanations, unifies disparate domains, and yields predictions that can be interrogated by experiment or mathematical derivation. Speculation thus becomes a *tool of falsification*, not an escape from it—it provides hypotheses broad enough to fail meaningfully.

In this light, the distinction between "scientific" and "speculative" is not binary but evolutionary. Every theory begins speculative; only its predictive success and empirical correspondence elevate it to orthodoxy. Newton's laws, Darwin's natural selection, Einstein's relativity—all began as thought experiments before they became cornerstones of knowledge. The responsible use of speculation, then, is not to assert truth prematurely, but to extend the reach of questioning itself.

By articulating frameworks like the Theory of Rendered Simulation (TORS), we are not abandoning scientific rigor; we are performing its highest function—the systematic exploration of alternative ontologies. Such frameworks remind us that science is not merely a catalog of results, but an evolving conversation about what counts as real, measurable, and meaningful.

To entertain the simulation framework, therefore, is not to drift into mysticism or metaphysics—it is to engage in *meta-physics* in the truest sense: the disciplined inquiry into what lies beneath the physical. It challenges us to view even the most fundamental assumptions of our models—continuity, locality, causality—not as immutable truths, but as potential renderings of a deeper informational substrate.

This kind of methodological pluralism is not a threat to science but its renewal. The history of discovery shows that when orthodox models fail to explain persistent anomalies, it is often the speculative frameworks—once dismissed as fanciful—that supply the conceptual leap required to move forward.

In short: speculation, when practiced with rigor, is not a violation of science. It *is* science in its most adventurous form. It is the act of building scaffolding before the cathedral is complete, of mapping possibilities before the landscape is fully known. And it is through this process—through frameworks like TORS—that we may glimpse the next horizon of understanding: where physics, computation, and consciousness converge into a single, unified ontology of reality.

Ontological Experimentation as Scientific Practice

Ontological experimentation is the deliberate and systematic exploration of the fundamental categories of being that structure not only our theories of the world but the very act of theorizing itself. It operates at the interface between metaphysics and empirical science, where the question is not merely *what exists* but *what it means for something to exist.* This form of inquiry does not seek to replace experiment with speculation but to reveal the hidden assumptions about time, causality, and substance that every experiment already presupposes.

In traditional science, ontology has often been backgrounded—treated as a settled framework within which measurements occur. Yet history shows that the greatest leaps forward have arisen precisely when those frameworks were questioned. When

Einstein reconceived simultaneity as relative, he was not only describing new data; he was redefining what time *is*. When quantum mechanics shattered the classical notion of locality, it demanded a rethinking of causation itself. Each of these transformations began as an ontological inquiry that, when formalized, produced empirically verifiable predictions. Ontological experimentation thus serves as the *pre-paradigmatic laboratory* where new conceptual scaffolds are built before they are tested in the empirical one.

In this sense, the practice of science is not purely empirical—it is also philosophical in structure, whether it admits it or not. Every hypothesis rests upon an implicit metaphysics: a belief about the continuity of space, the linearity of time, or the independence of observer and observed. To experiment ontologically is to bring these implicit metaphysics into view, to manipulate them, and to ask whether alternative metaphysical commitments yield a more coherent or predictive picture of the cosmos.

Incorporating ontological experimentation into scientific method means treating conceptual innovation as a legitimate form of data generation. A new ontology, when internally consistent and heuristically powerful, can be as revelatory as a new observation. For instance, simulation theory reframes the ontological substrate of physics from material substance to computational process—an act of conceptual engineering that generates novel testable questions about discreteness, information limits, and the architecture of physical law. Similarly, panpsychism and integrated information theory reinterpret consciousness as a fundamental aspect of reality rather than an emergent one, thereby reshaping the explanatory boundary between physics and phenomenology.

By explicitly engaging in ontological experimentation, scientists and philosophers become collaborators in the same enterprise: the discovery of *what it means for something to be real*. This approach demands both rigor and imagination—a willingness to subject metaphysical models to the same scrutiny we apply to mathematical ones, and an openness to the possibility that "reality" itself may be a variable, not a constant.

In the coming era, as physics, computation, and consciousness studies converge, ontological experimentation may become not just a theoretical exercise but a methodological necessity. It will serve as the compass guiding inquiry into domains where observation alone cannot reach—quantum indeterminacy, spacetime discreteness, or the simulated substrate of existence. To experiment ontologically is to recognize that science's ultimate frontier is not the discovery of new particles or galaxies, but the discovery of new ways to conceive of what *is*.

Criteria for Productive Speculation

Speculation becomes scientifically meaningful when it adheres to a disciplined methodology—when it is not an escape from rigor but an extension of it into uncharted conceptual terrain. In this light, productive speculation functions as a form of theoretical prototyping: it allows for exploratory thought that remains grounded in coherence, empirical potential, and intellectual humility. The following criteria serve as the scaffolding for speculation that advances rather than undermines scientific understanding.

1. Logical Coherence

A speculative framework must first be internally consistent. Its premises cannot contradict one another, and its conclusions must follow logically from those premises. Logical coherence does not guarantee truth, but it is the minimal condition for intelligibility. Without it, speculation devolves into narrative or mysticism. This standard ensures that even highly imaginative ideas—such as the simulation hypothesis or multiverse cosmology—retain a structure that can be examined, debated, and potentially refined through reasoned analysis. Logical coherence is the grammar of disciplined imagination; it allows speculation to be taken seriously as philosophy and tested seriously as science.

2. Cross-Disciplinary Integration

The most powerful speculative frameworks are those that bridge domains rather than isolate them. Reality is not compartmentalized, and neither should be our inquiry into it. A framework that proposes, for instance, that spacetime is emergent from quantum information must also cohere with thermodynamics, cosmology, and even cognitive science, since the human act of observation participates in how reality manifests. Integration across disciplines ensures that speculative ideas are stress-tested from multiple directions. A theory that collapses under interdisciplinary scrutiny reveals itself as parochial—a curiosity rather than a foundation. Conversely, frameworks that resonate across fields often point toward deeper structural truths, suggesting that diverse phenomena share an underlying informational or computational logic.

3. Generativity

The worth of speculation lies not in its certainty but in its fertility—the extent to which it generates new questions, models, or avenues for empirical testing. A productive speculative framework acts as a conceptual seedbed from which unexpected insights grow. Einstein's thought experiments, Schrödinger's cat, and Turing's machine were speculative constructs, yet each catalyzed revolutions in understanding by reframing what questions could even be asked. Generativity transforms speculation from idle conjecture into a creative engine of discovery. If an idea provokes new methods of measurement, or reinterprets known data in a novel way, it fulfills its scientific function even before it is proven.

4. Falsifiability in Principle

A framework need not be testable *immediately* to be scientific, but it must be *testable in principle*. There must exist some conceivable state of evidence—some configuration of observations—that could verify or refute it. This condition preserves the self-correcting nature of science. For instance, even if direct access to the "code" of a simulated universe is impossible, the hypothesis can still be constrained by searching for signatures of digital discreteness, computational limits, or non-random informational patterns in quantum phenomena. The criterion of falsifiability disciplines the imagination; it forces speculation to remain anchored to a world in which evidence matters.

5. Epistemic Humility

Finally, productive speculation must be accompanied by an awareness of its provisional nature. To speculate responsibly is to understand that one's framework is not a revelation but a proposal—a lens, not the landscape. Epistemic humility guards against dogmatism and reminds us that every model, however elegant, is a simplification of something greater than human cognition can fully grasp. This humility does not weaken the speculative enterprise; it strengthens it by preserving openness to revision, reinterpretation, and surprise. The most enduring theories in science are those that anticipate their own evolution.

In Synthesis

Together, these criteria delineate a form of speculation that is neither naive nor nihilistic. It acknowledges the necessity of imagination while maintaining allegiance to evidence and logic. Productive speculation serves as the *bridge between the known and the conceivable*—a disciplined act of creative reasoning that expands the map of inquiry without severing it from the terrain of reality. When speculation meets these standards, it ceases to be a detour from science and becomes one of its most vital engines of progress.

Why This Belongs in Science

To reintroduce ontological and speculative inquiry into science is not to blur the line between philosophy and empiricism, but to restore a relationship that was never meant to be severed. Science, in its deepest sense, is not the mere accumulation of data or the perfection of technique—it is a systematic attempt to understand *what reality is and how it comes to be known.* Every great scientific revolution has involved not only new instruments or equations, but a redefinition of what counts as real, measurable, and possible.

When Galileo turned his telescope to the heavens, he was not simply collecting optical data; he was overturning the Aristotelian ontology of a perfect celestial sphere. When Newton introduced gravity as an invisible force acting at a distance, he expanded the boundaries of what "action" meant in nature. When Einstein reimagined space and time as dynamic rather than absolute, he did not just refine celestial mechanics—he redefined the very architecture of existence. Each of these breakthroughs was, at heart, an ontological experiment: a test of the assumptions that underwrite reality itself.

To deny ontological exploration today is to freeze science at the level of description—to let it catalog phenomena without ever reexamining the deeper scaffolding of being. This risks turning science into a museum of facts rather than a living inquiry. Yet science has never advanced by staying within its comfort zone; it evolves precisely when it confronts the inadequacy of its current categories. Quantum mechanics, relativity, and information theory all began as speculative frameworks that dared to challenge the metaphysical foundations of their time.

The simulation hypothesis, along with other emerging frameworks in digital physics and information-based ontology, continues this tradition. It asks not merely *how* physical processes unfold, but *why* the universe behaves as if it were governed by rules that resemble computation—discrete, structured, and optimally efficient. Such questions are not distractions from empirical work; they provide new avenues for it. The idea that spacetime might be quantized or that entropy could be an informational property did not arise from data alone, but from conceptual reexamination of what "space," "energy," and "information" mean.

Incorporating ontological experimentation into the scientific enterprise also acknowledges that science operates within a conceptual horizon—it is a way of *seeing*, not an unmediated access to truth. To refine that horizon is to refine science itself. Asking whether time is fundamental or emergent, whether consciousness is a local computation or a non-local phenomenon, whether physical law is written or evolved—these are not idle speculations. They are extensions of the scientific spirit: the relentless drive to know not only *what* reality does, but *what it is*.

Ultimately, this approach restores balance between empiricism and imagination. Data without ontology risks becoming inert—a flood of observations without meaning. Ontology without data risks collapsing into metaphysics—a structure without grounding. But together, they form the dynamic tension that powers genuine discovery.

Science belongs not only in the laboratory, but in the question that precedes every experiment: *What counts as real?* To exclude ontological inquiry from science would be to amputate the very limb that allows it to reach beyond the known. In embracing it, science reclaims its full identity—not as a closed system of knowledge, but as a living, evolving dialogue between what is observed and what is possible.

About the Author

Robert Restivo is not a physicist by training — and that is precisely the point. Holding degrees in law, business, and nursing, he has spent his life navigating the very human worlds of systems, structures, and crises. From boardrooms to emergency rooms, he has seen how rules — whether legal codes, economic incentives, or biological imperatives — shape behavior, often invisibly, until something breaks.

But it is precisely this multidisciplinary path that has shaped his thinking: rigorous argument from law, systems-level analysis from business, and a frontline intimacy with human fragility from medicine.

Though not a physicist by training, Robert approaches questions of existence with the curiosity of an outsider unburdened by orthodoxy. For years, he has been drawn to the simulation hypothesis as a framework that unifies what science, philosophy, and religion each struggle to explain in isolation.

That perspective colors his approach to physics and metaphysics. A lifelong tech enthusiast, Restivo has always been drawn to the question of whether reality itself might be governed by deeper "code" — rules of a higher order, hidden beneath the appearance of matter, time, and space. His study of philosophy, combined with his clinical eye for anomalies and systems that don't quite add up, led him toward the Simulation Hypothesis as the most powerful unifying framework for both science and religion.

Beyond the page, Robert is a lifelong technophile, fluent enough in computers to suspect they may be whispering deeper truths about our own "hardware" and "software." He speaks German and Italian with the fluency of an ambitious toddler, owns a guitar he still swears he'll learn to play, and spends as much time as possible outdoors with his dogs.

Above all, he believes humanity can — and must — reach for a future worthy of its imagination. Whether that future is hard-coded into the universe or dynamically rendered in real time, Robert is determined to help us see the pattern before the screen goes dark.

Made in the USA
Monee, IL
12 November 2026

ae0ef73f-8d1e-43d1-a219-961fc6b984b9R01